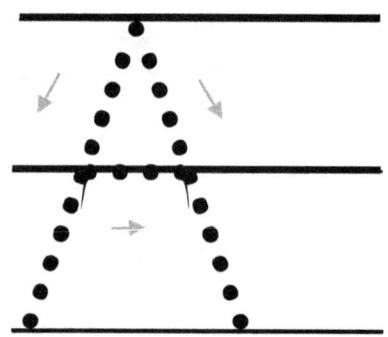

alligator

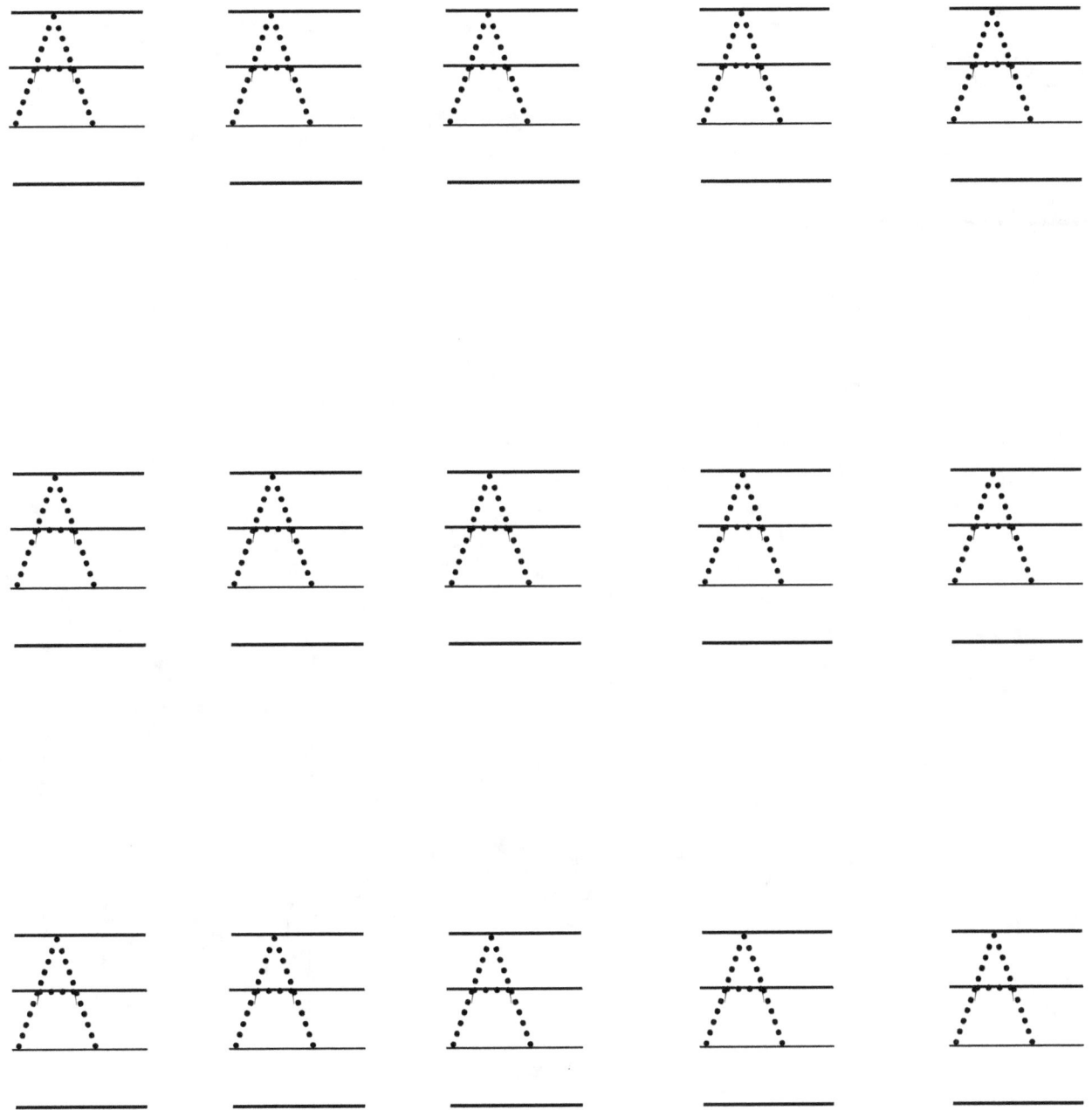

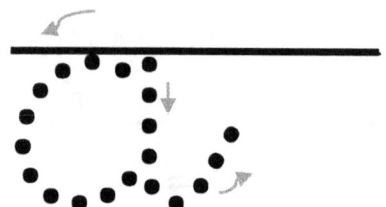

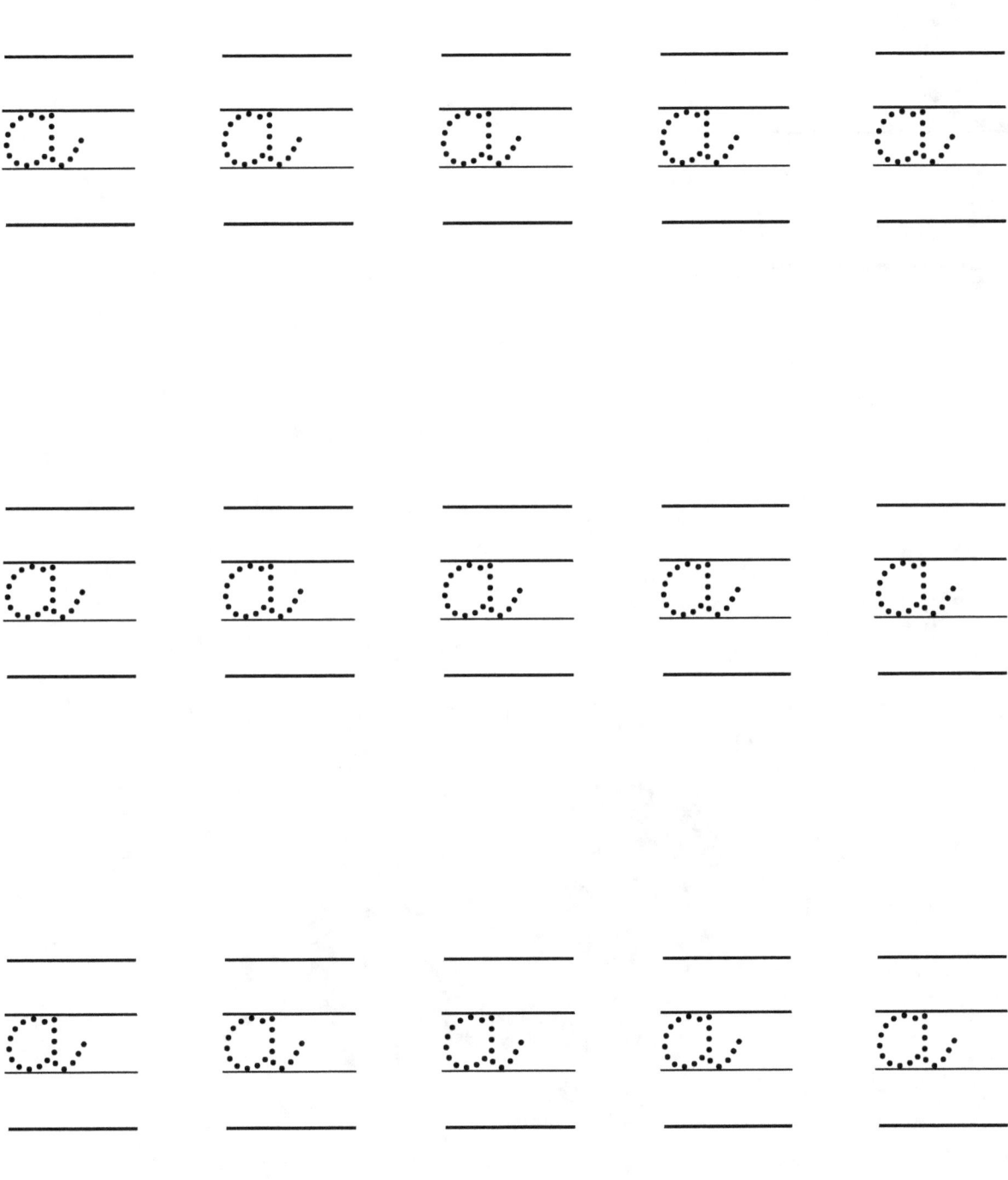

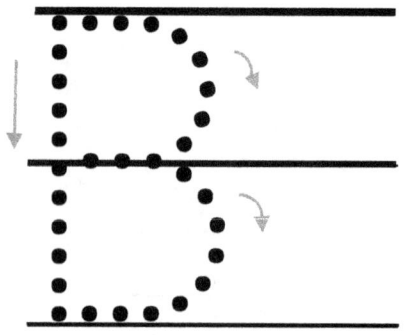

bee

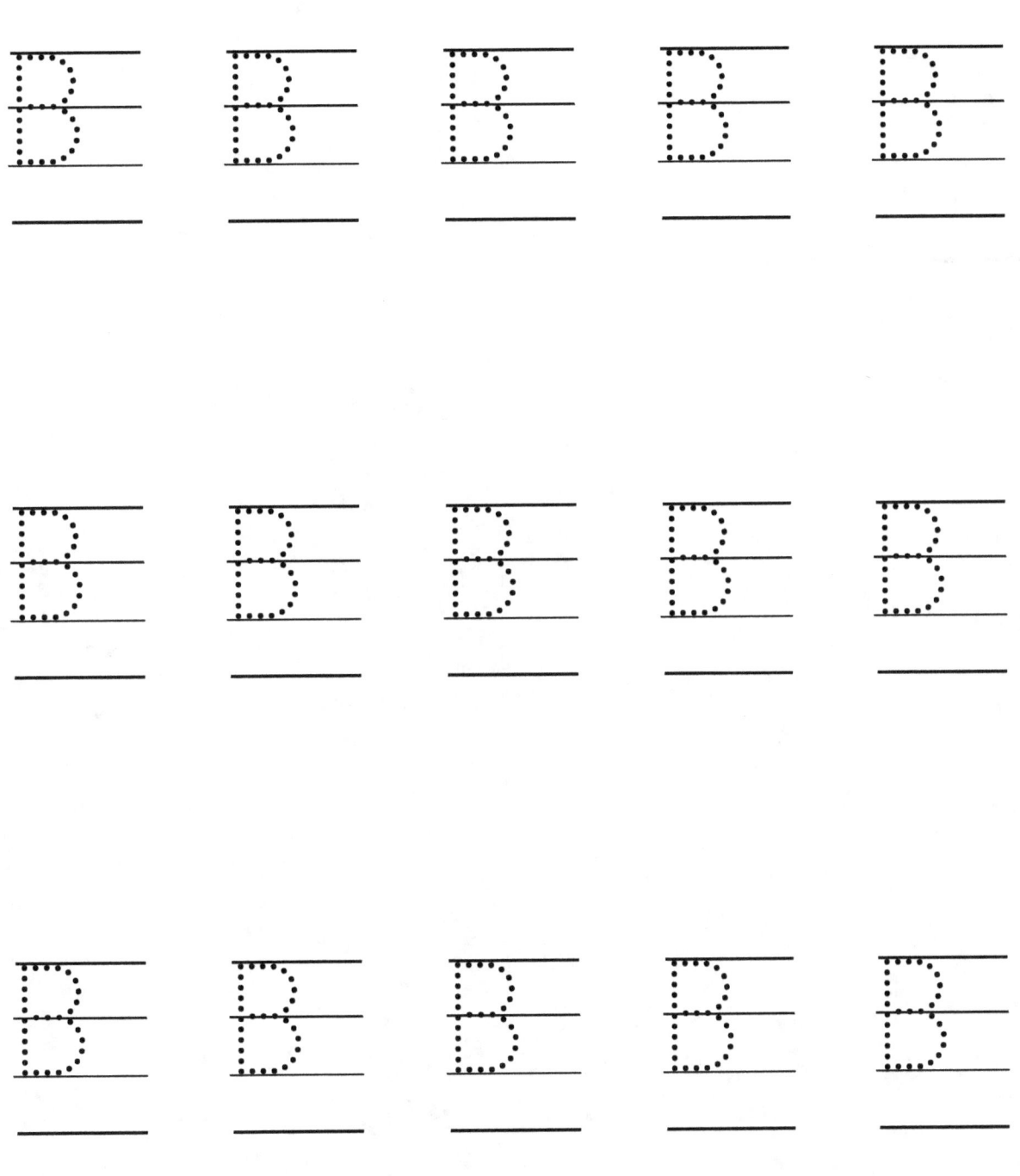

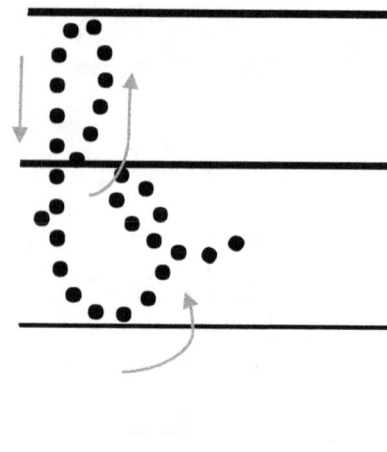

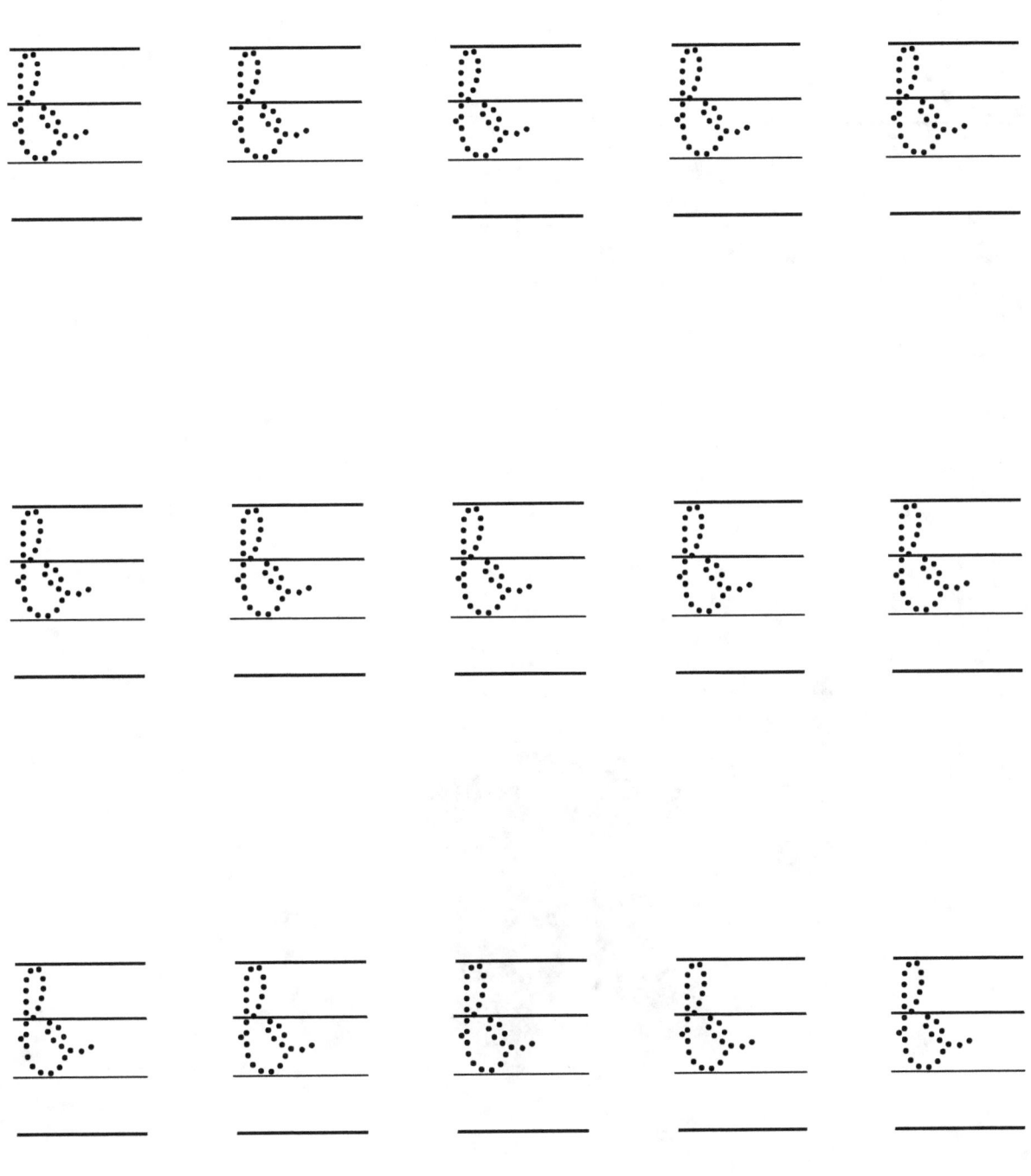

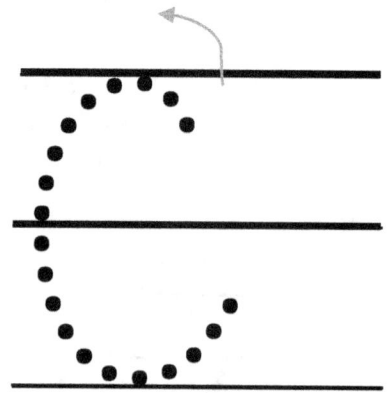

cat

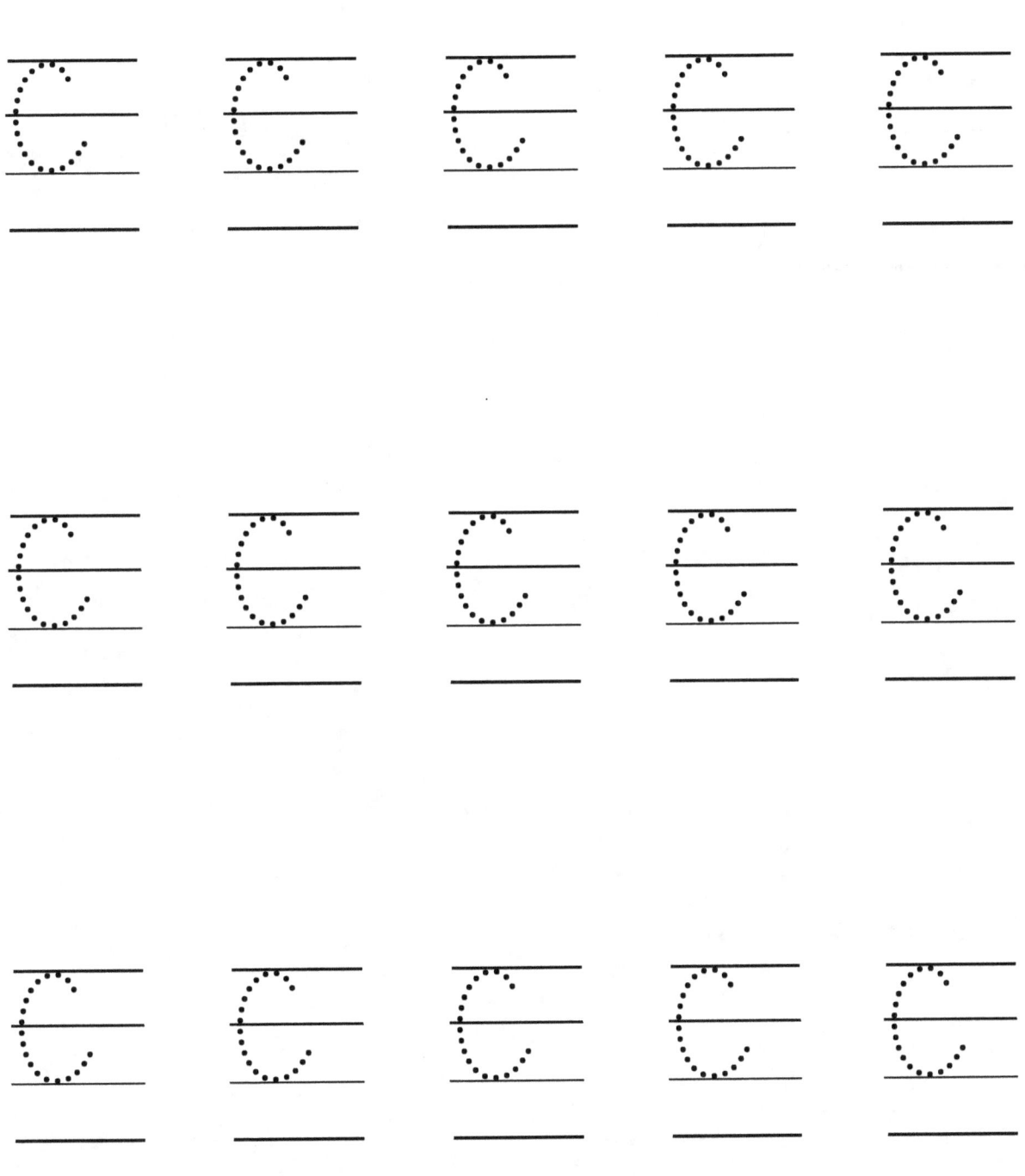

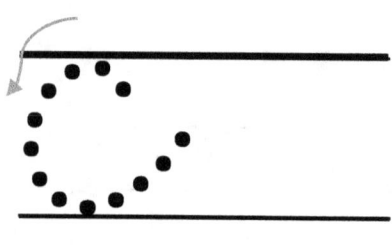

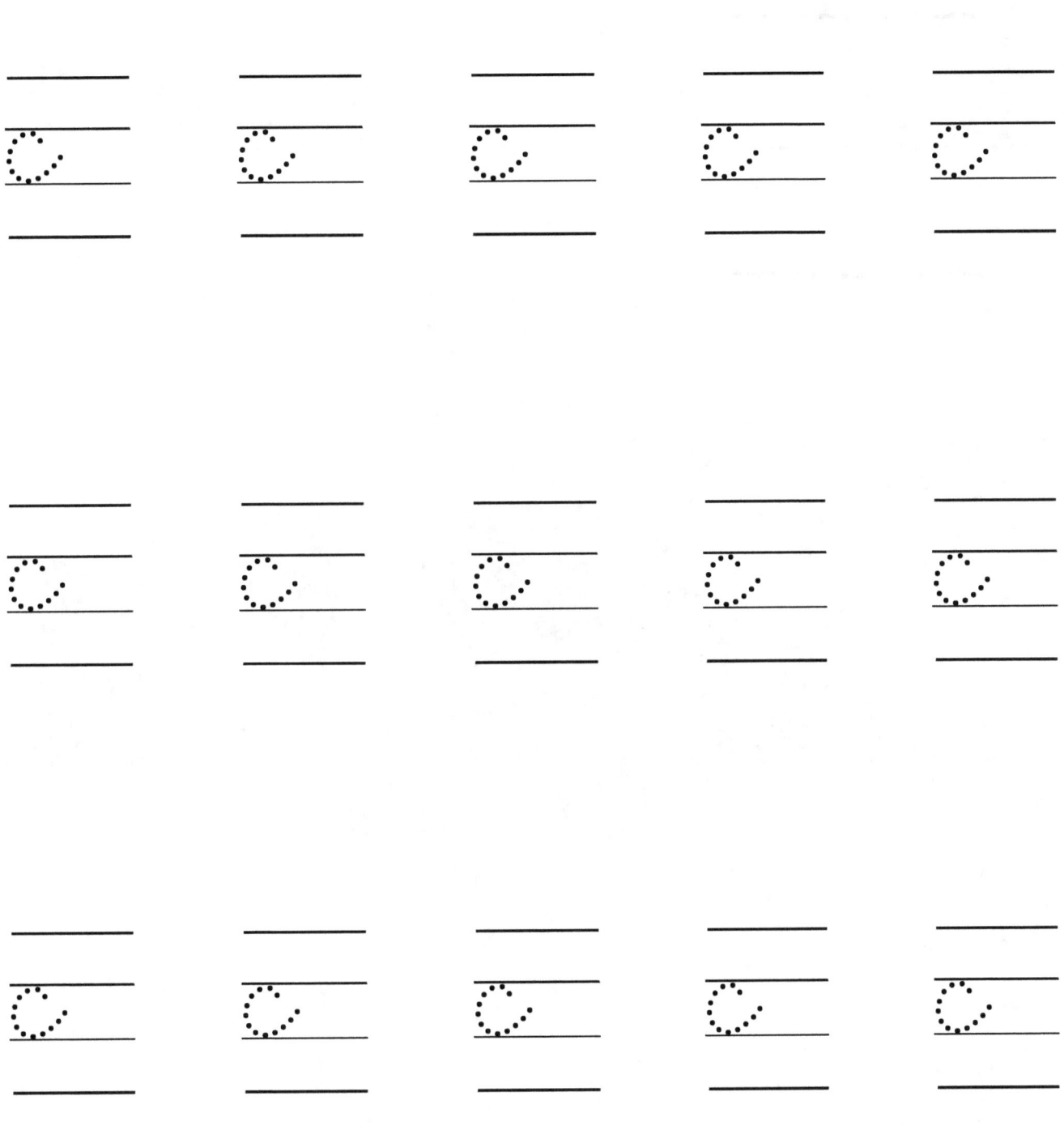

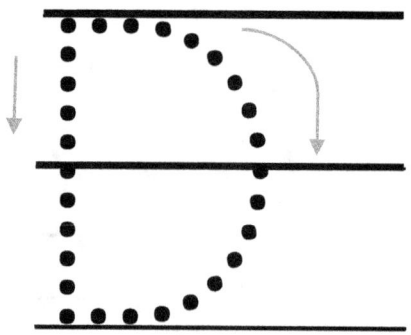

dog

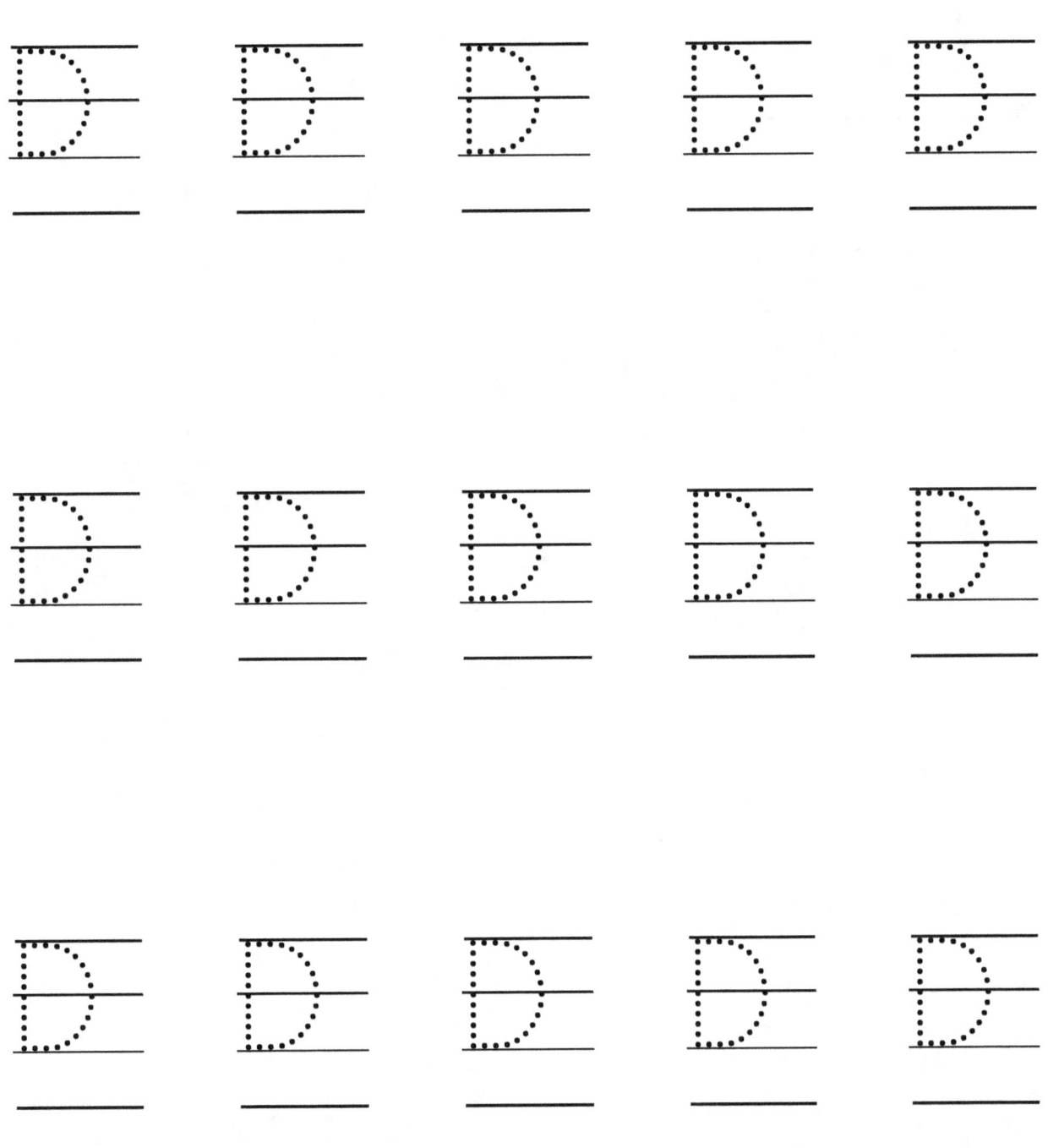

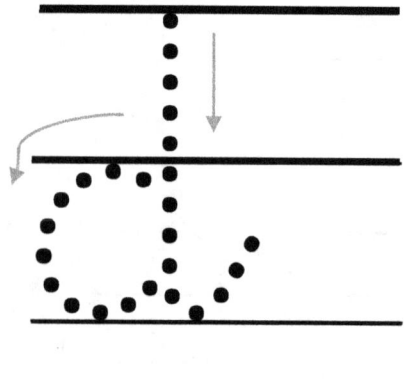

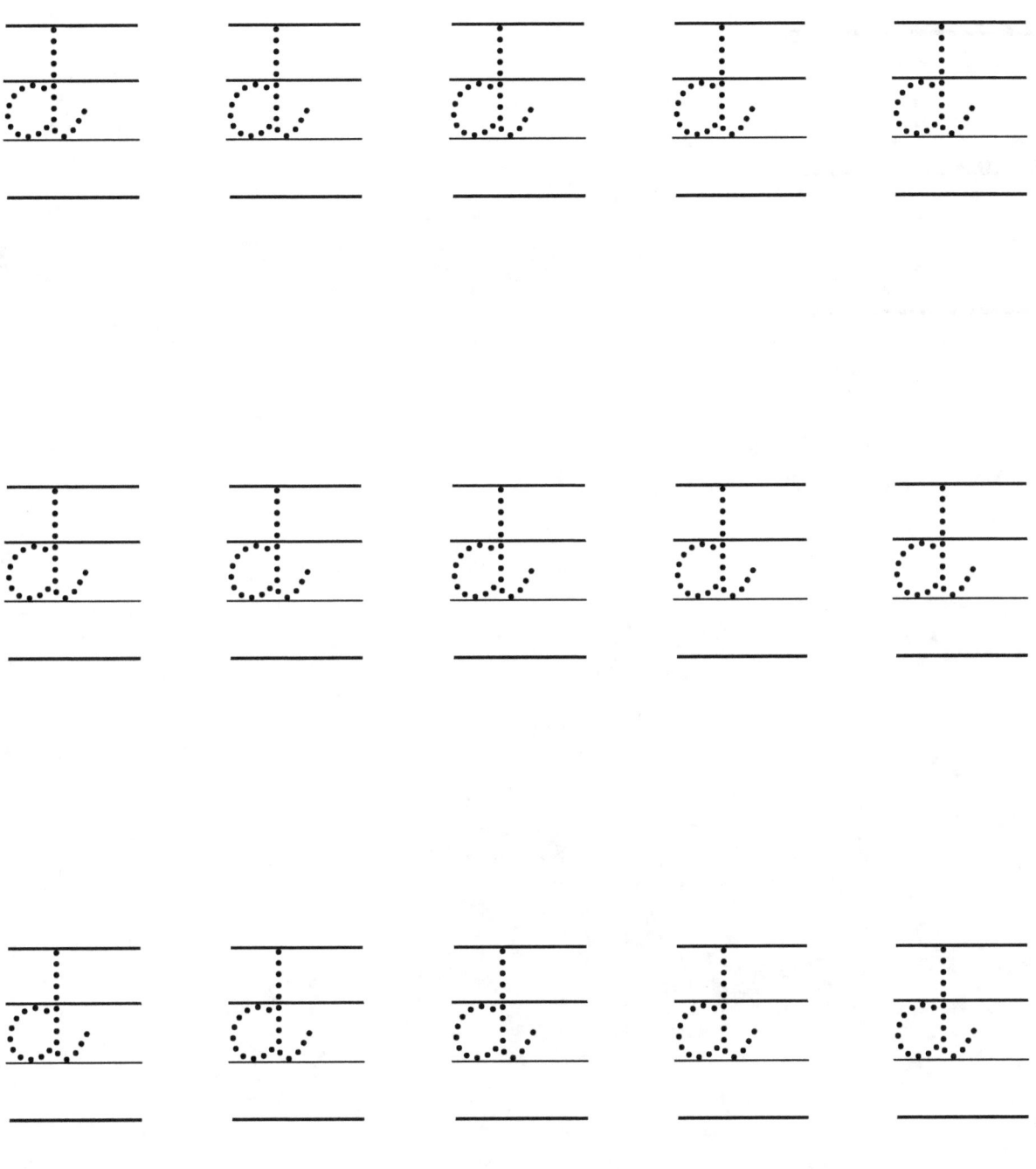

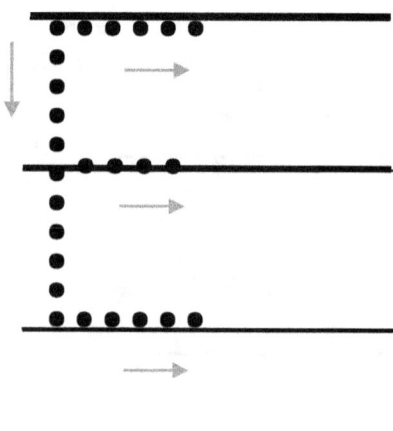

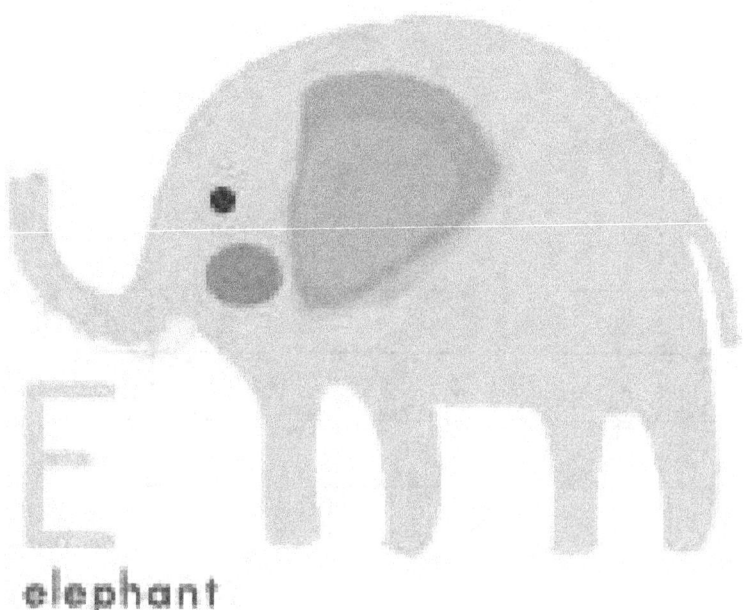

elephant

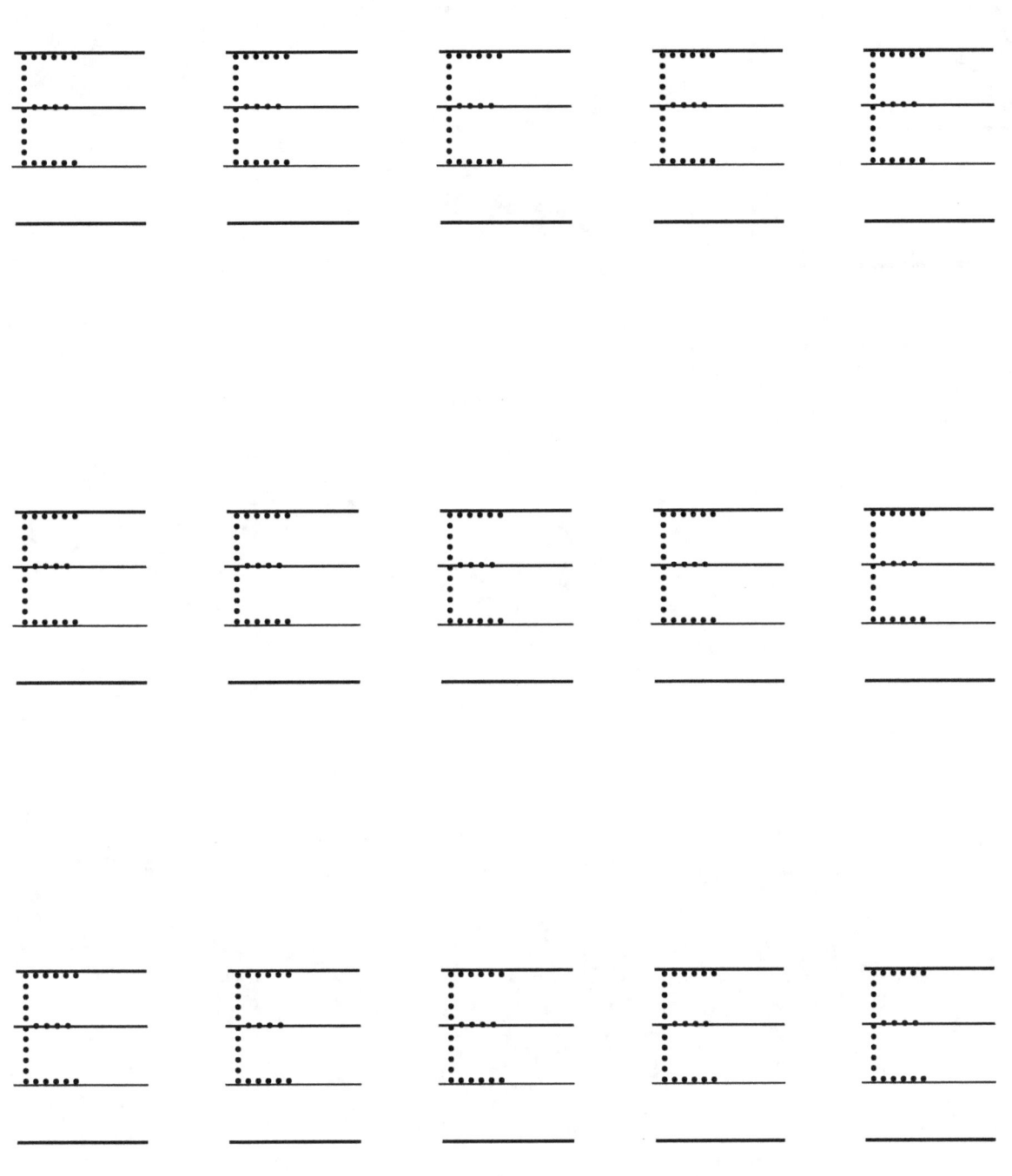

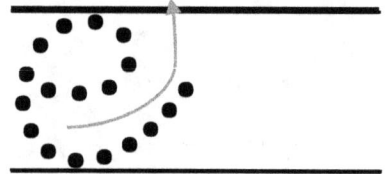

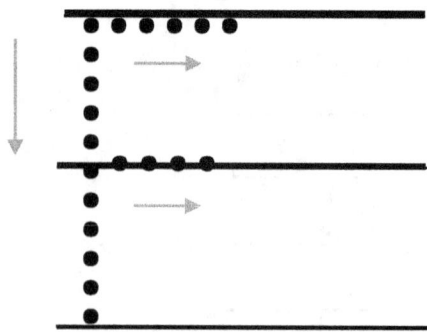

fox

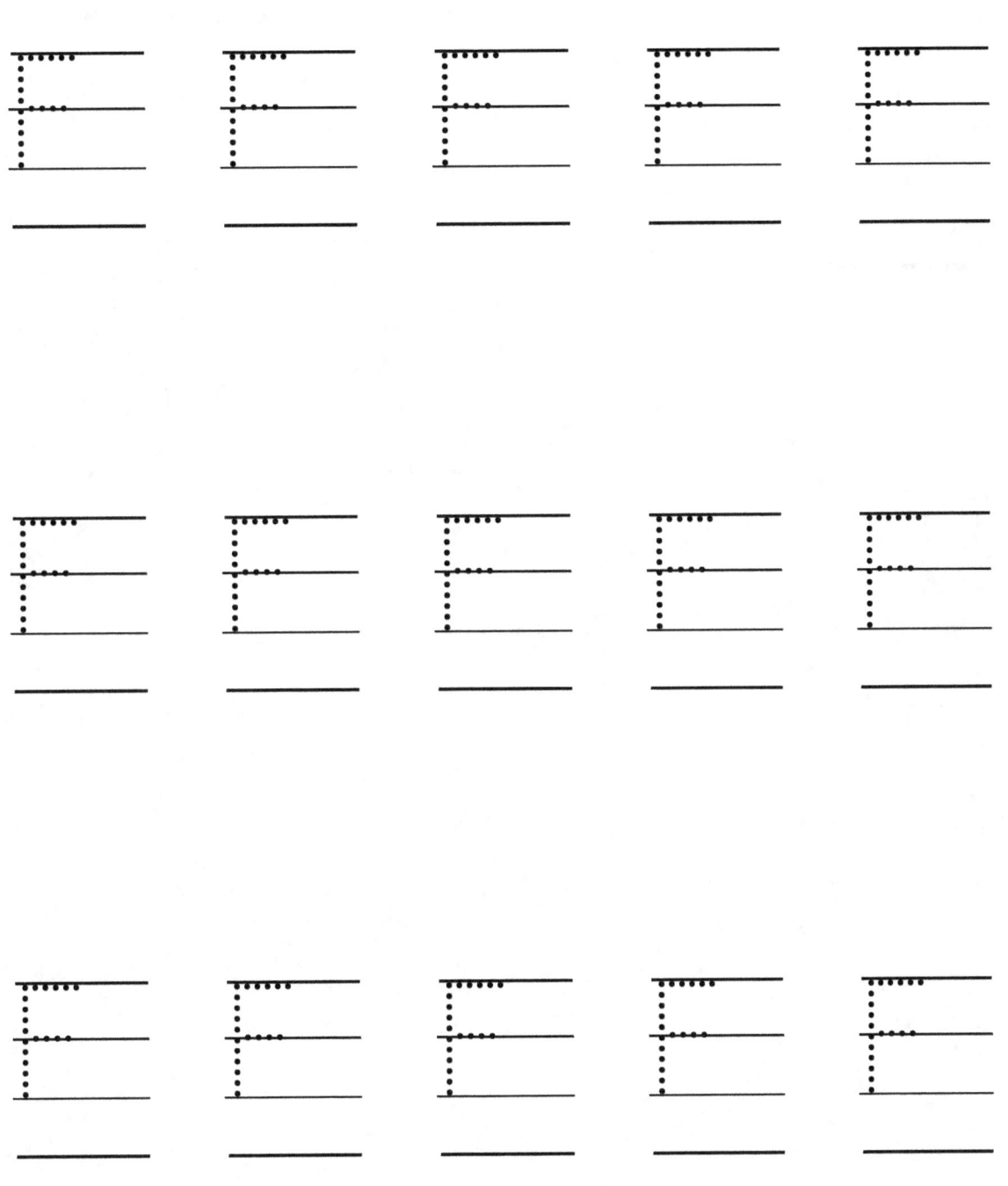

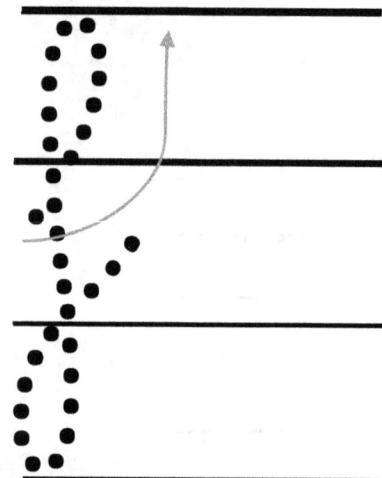

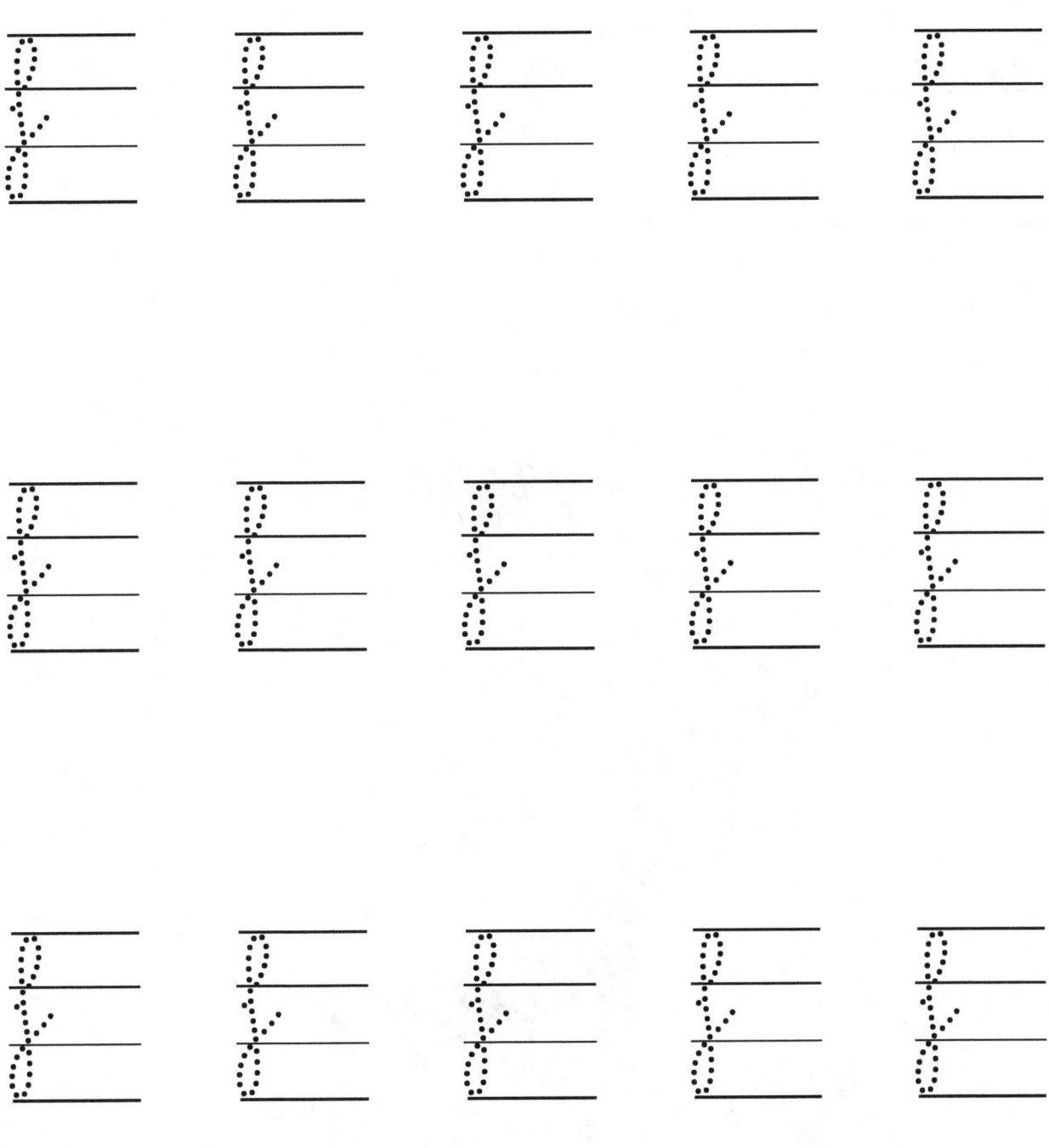

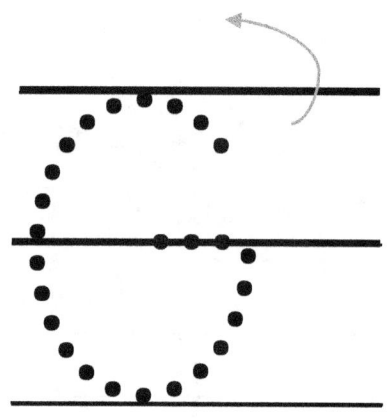

G *giraffe*

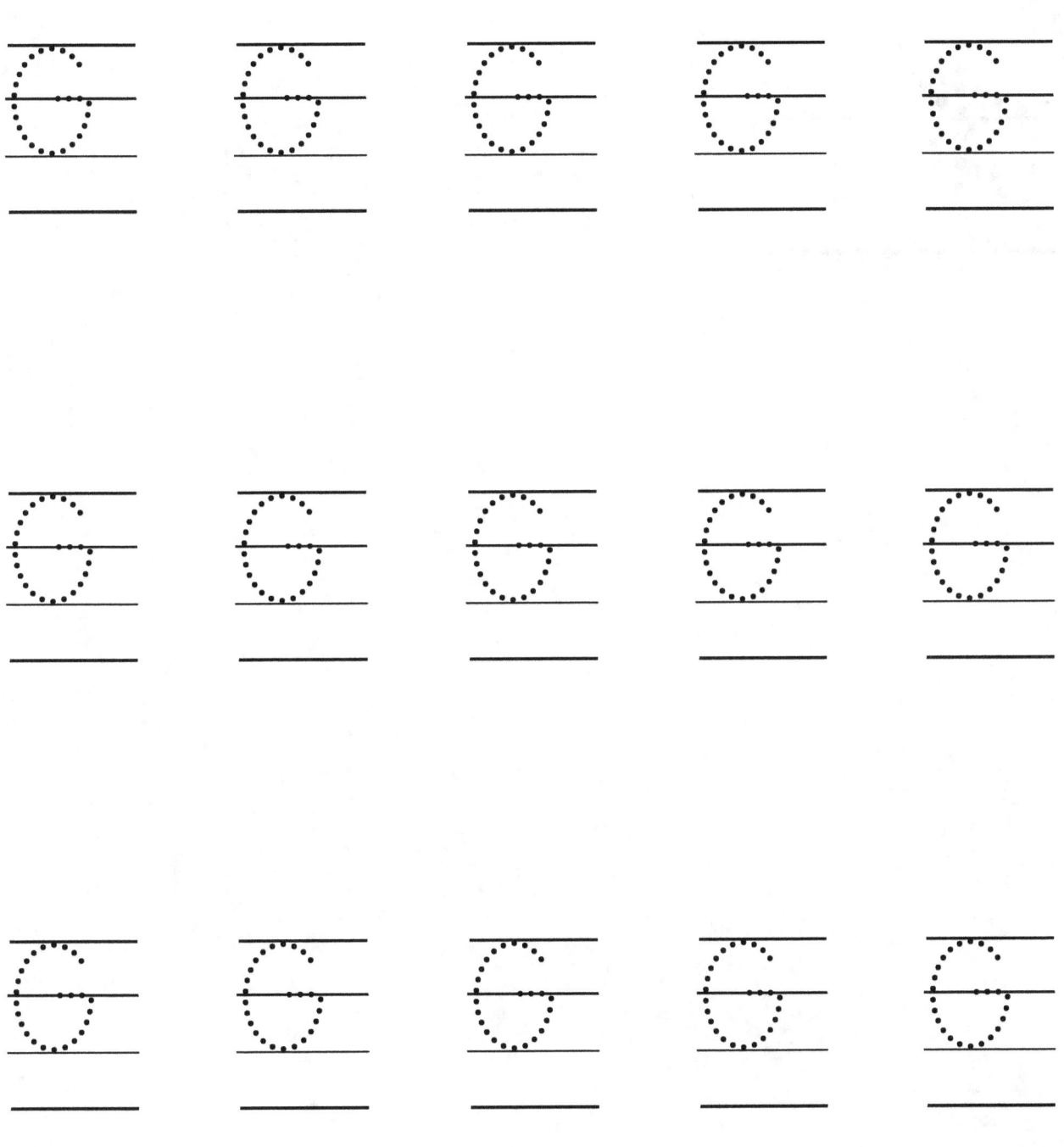

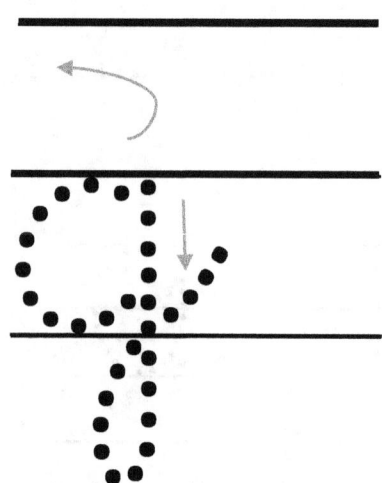

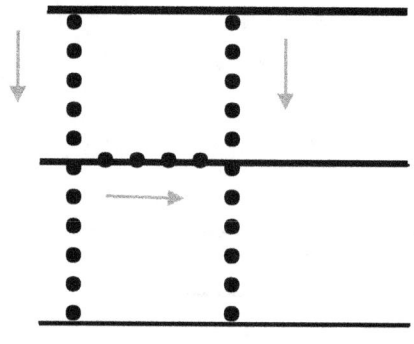

hippo

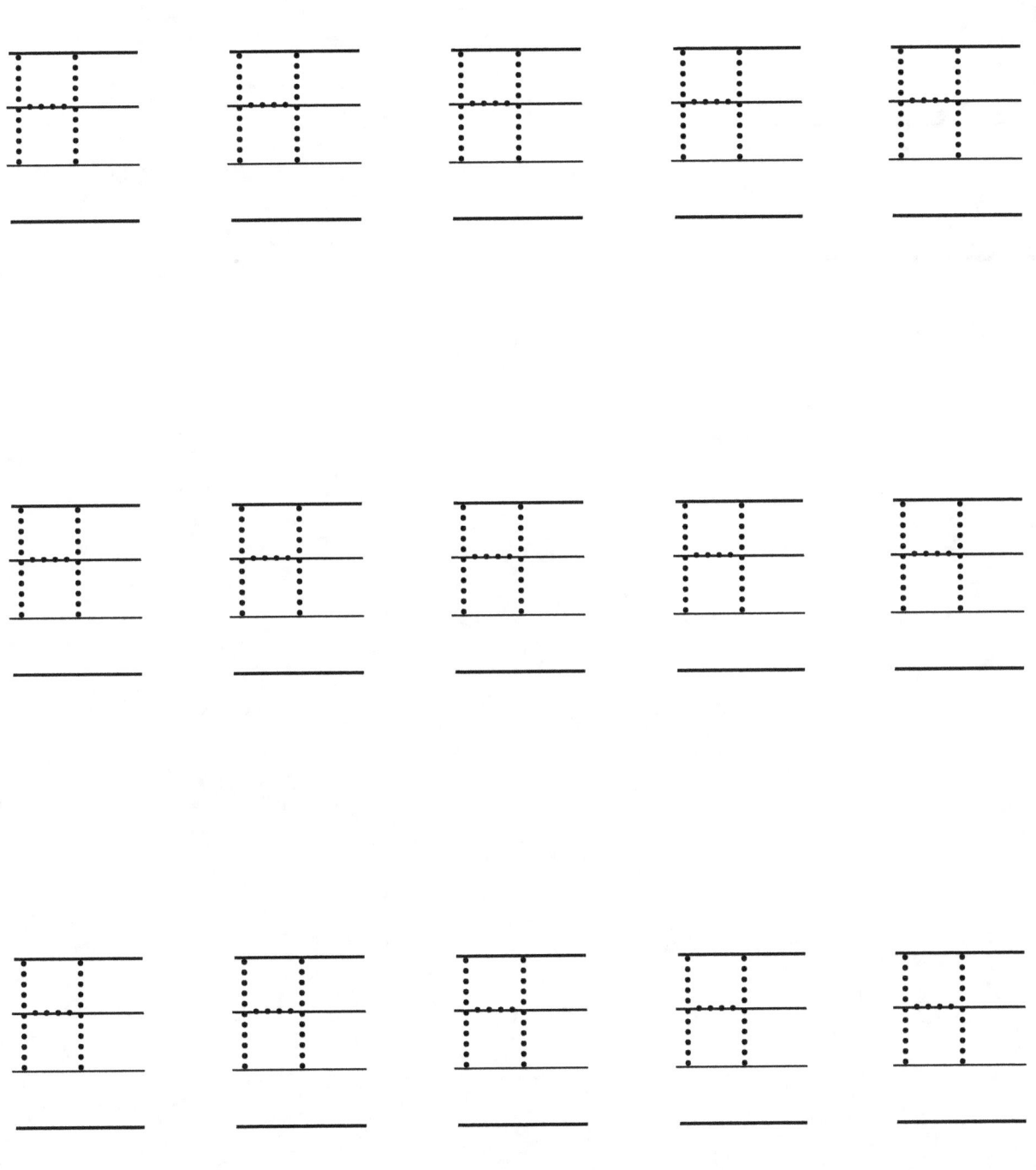

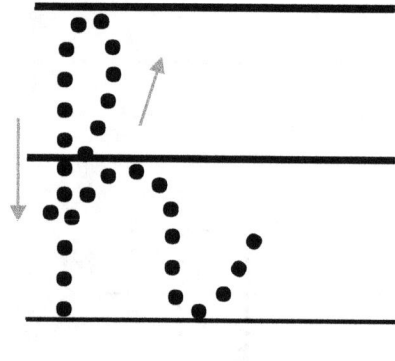

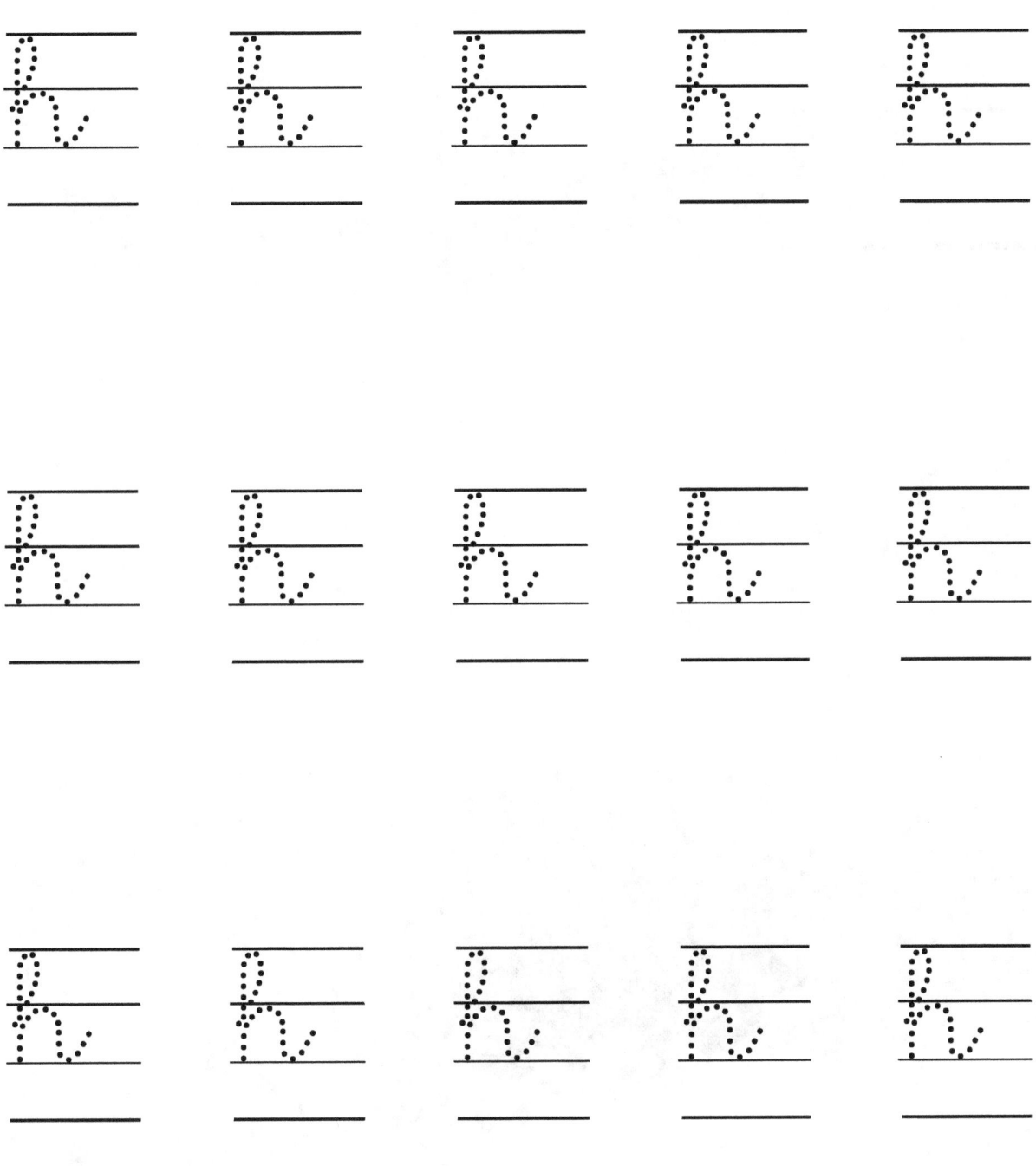

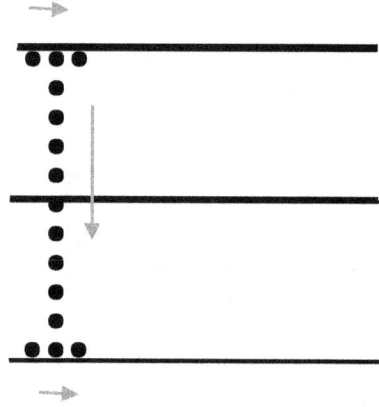

iguana

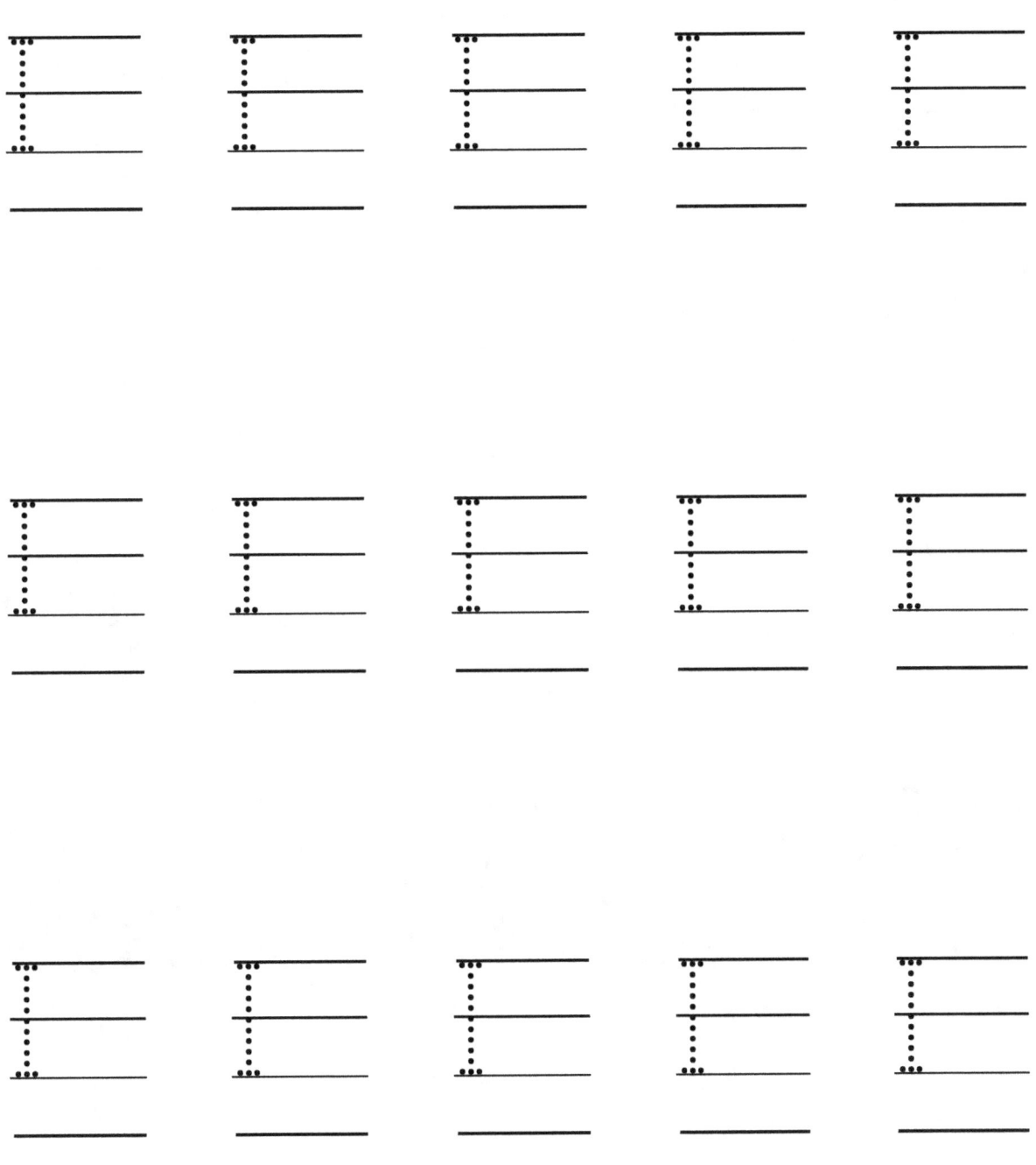

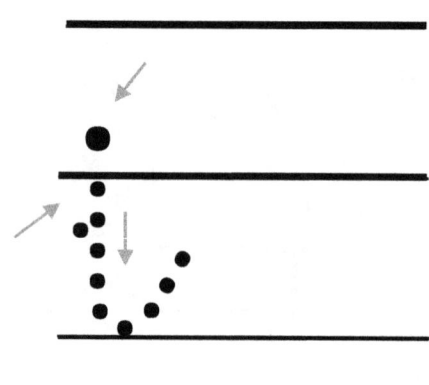

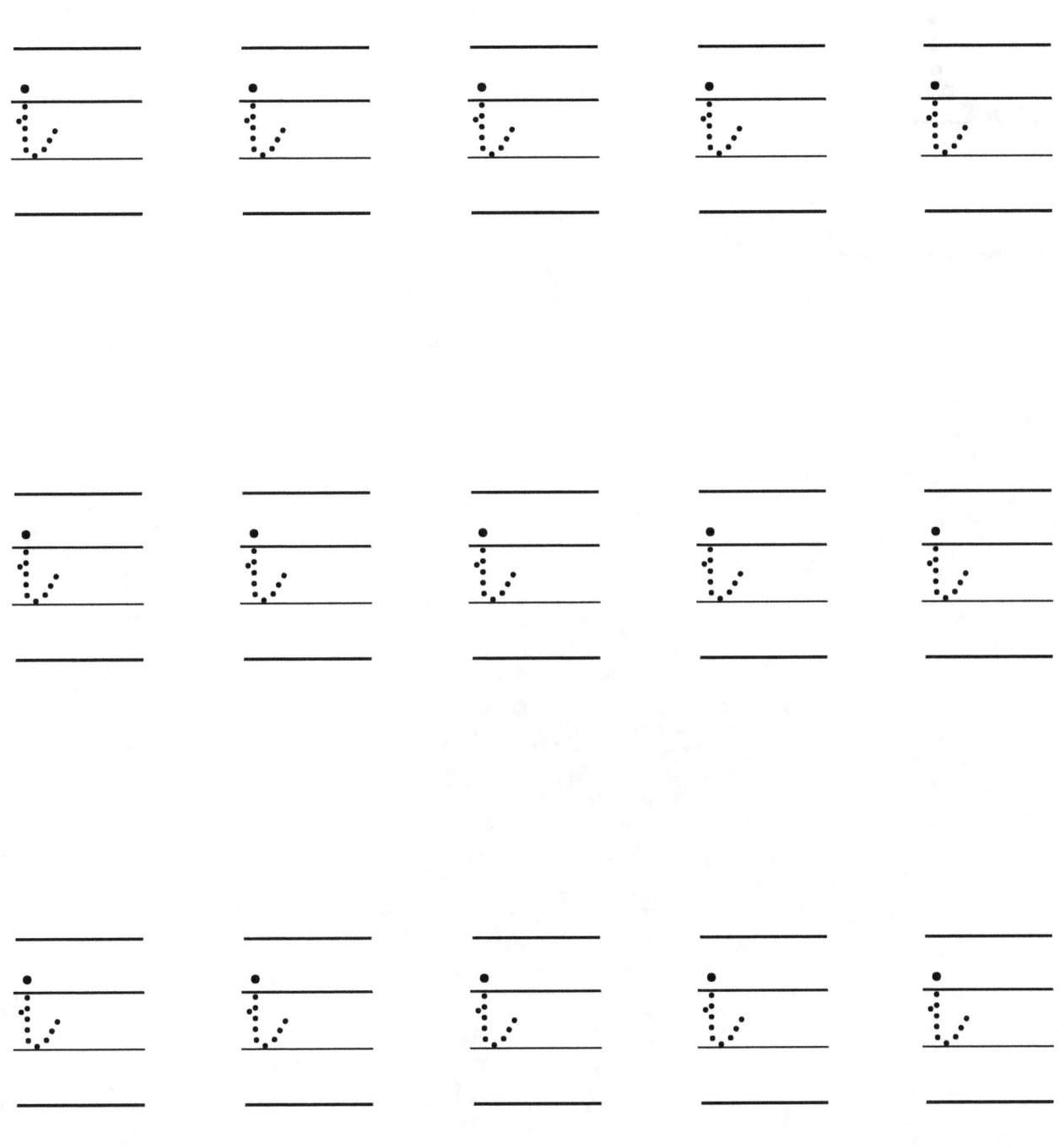

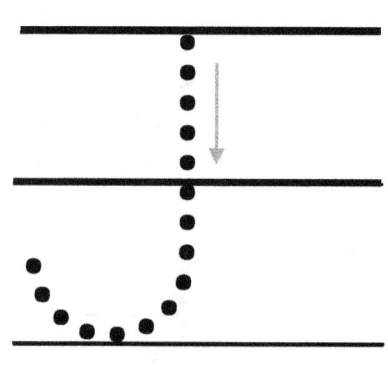

jellyfish

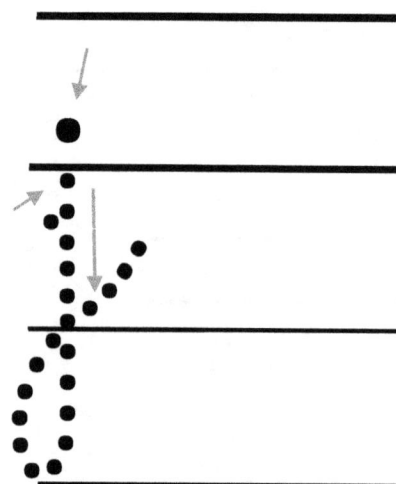

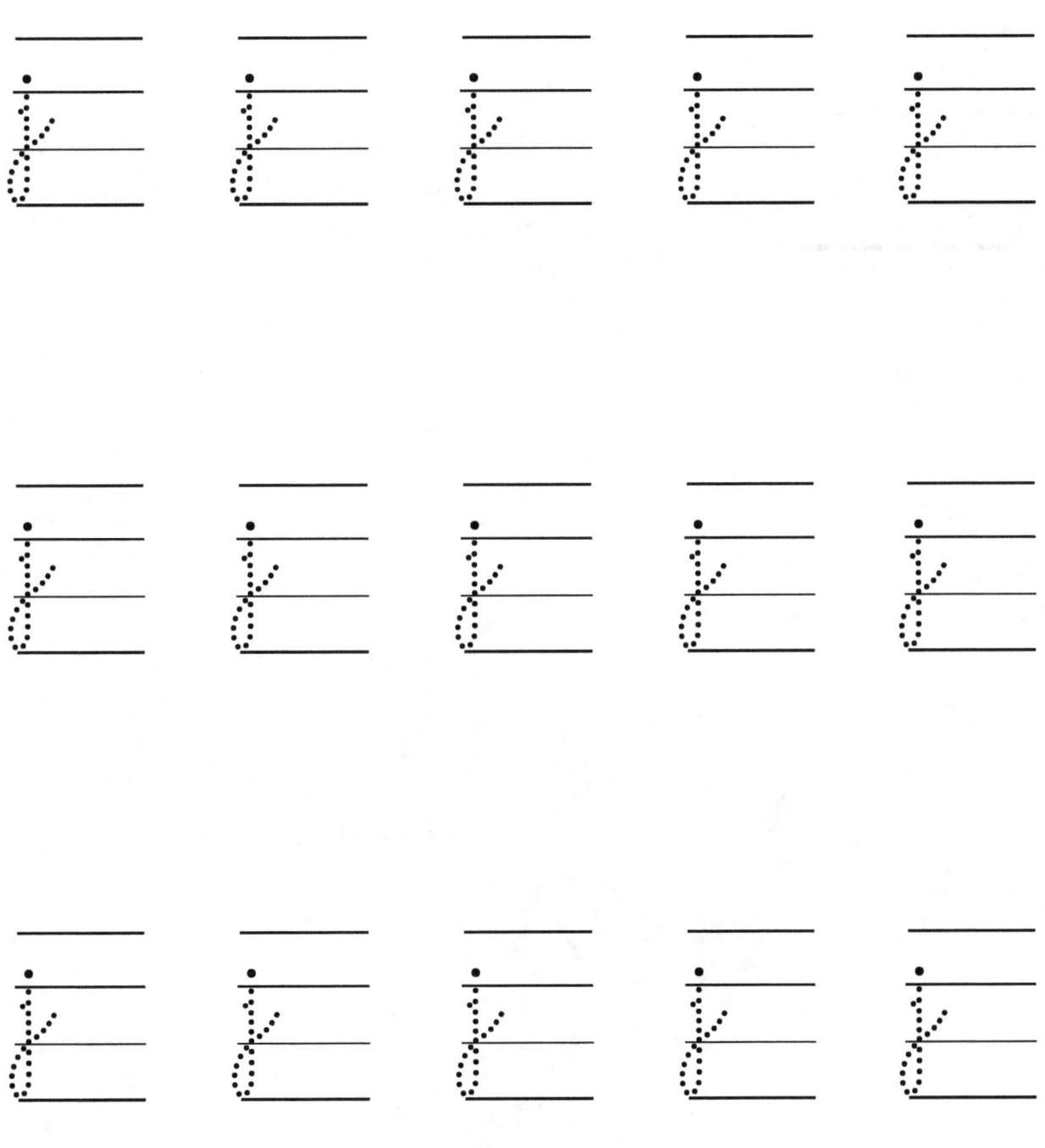

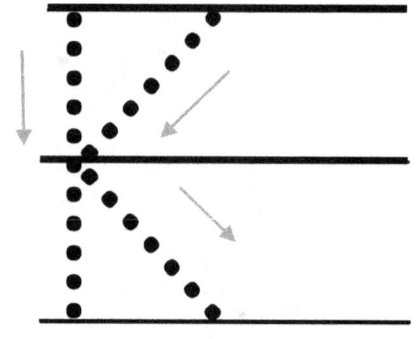

K
kangaroo

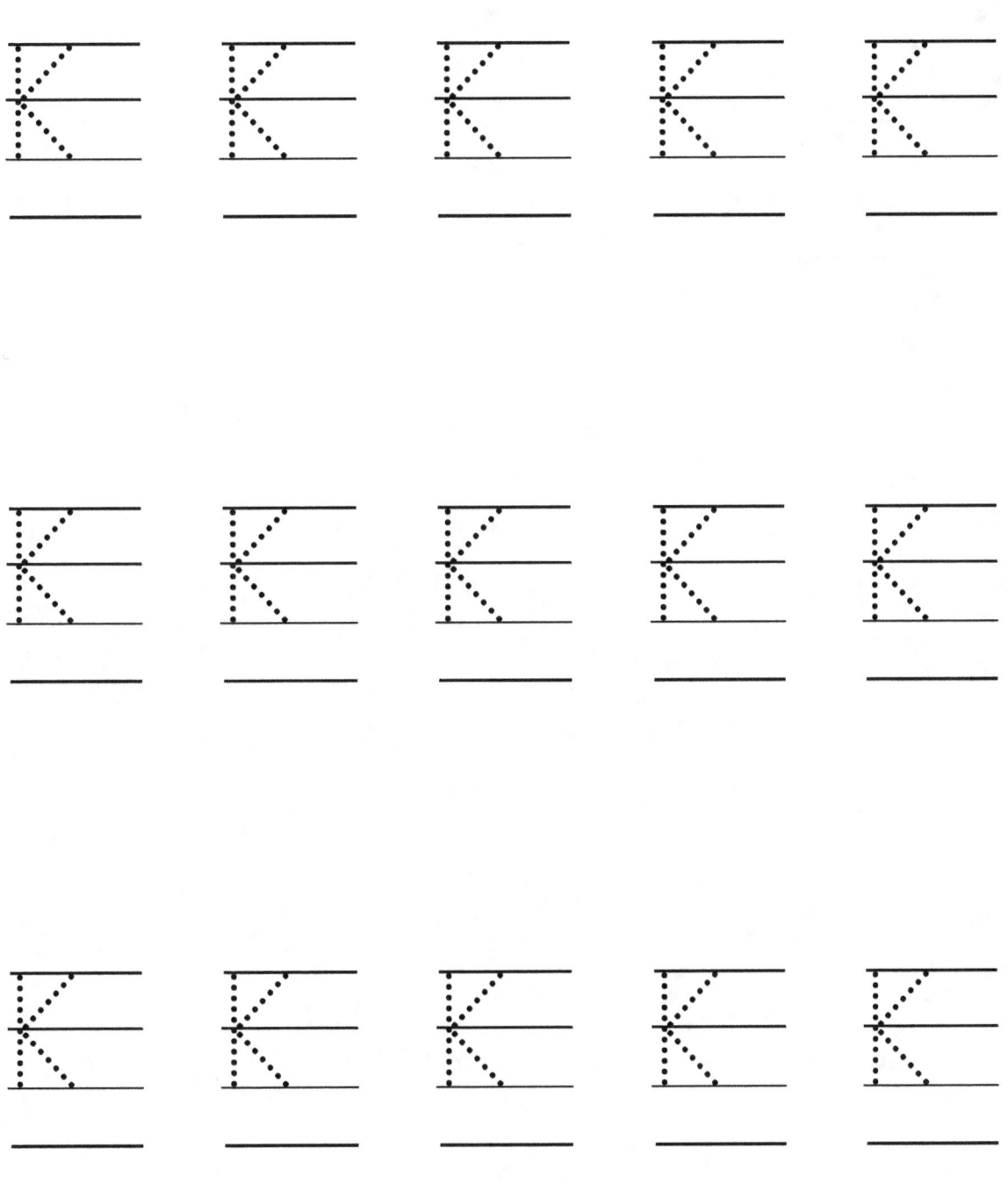

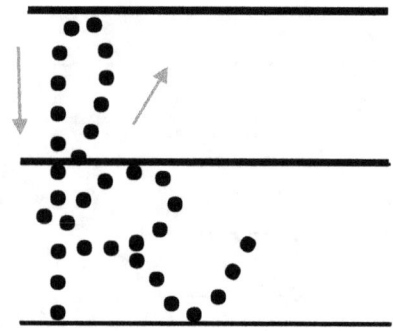

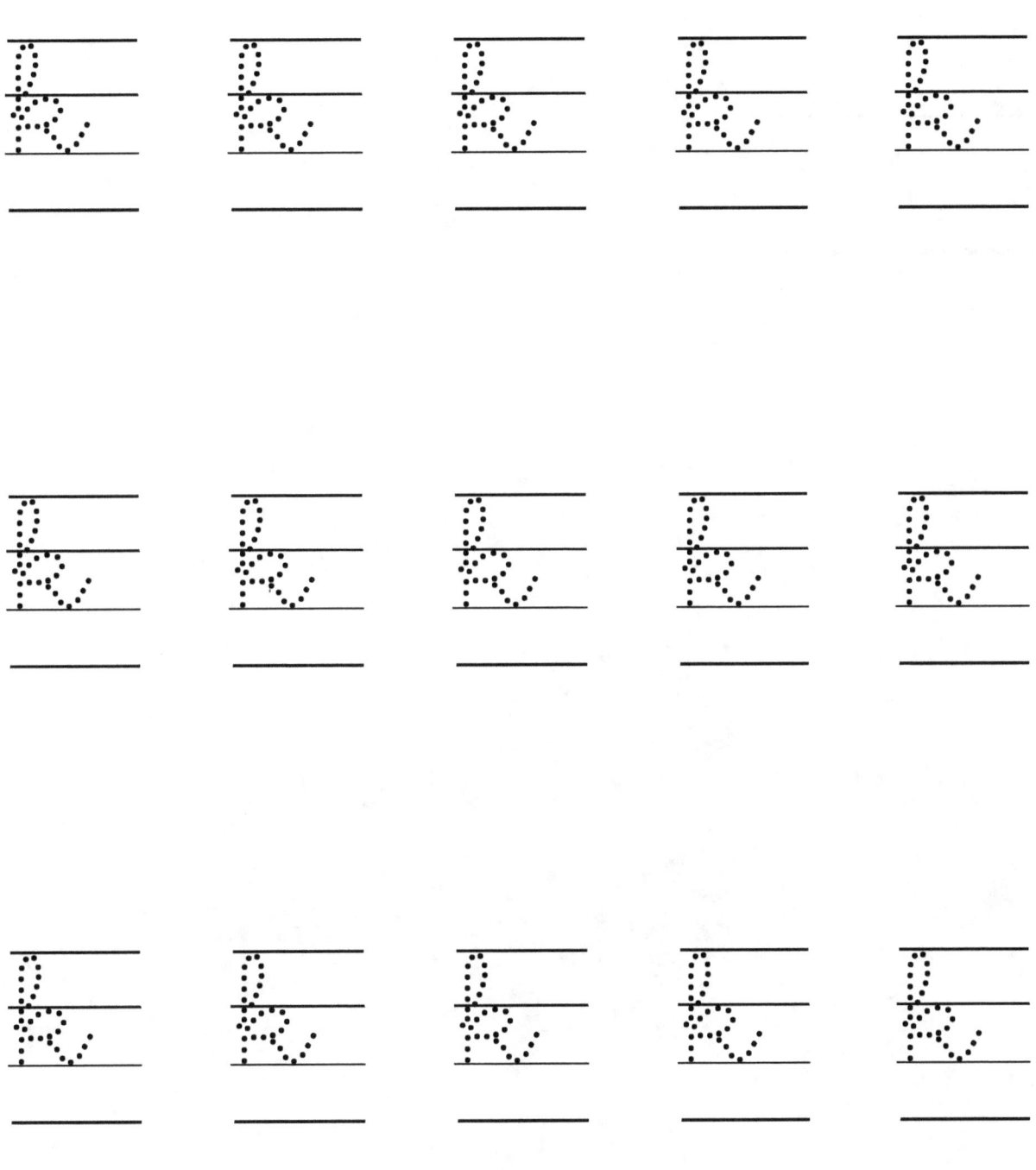

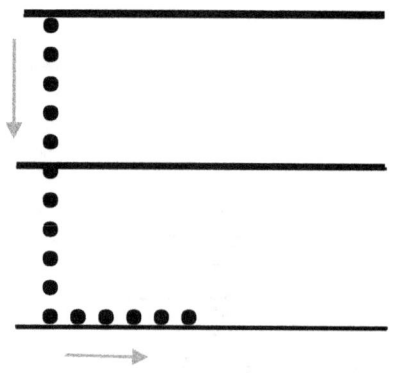

lion

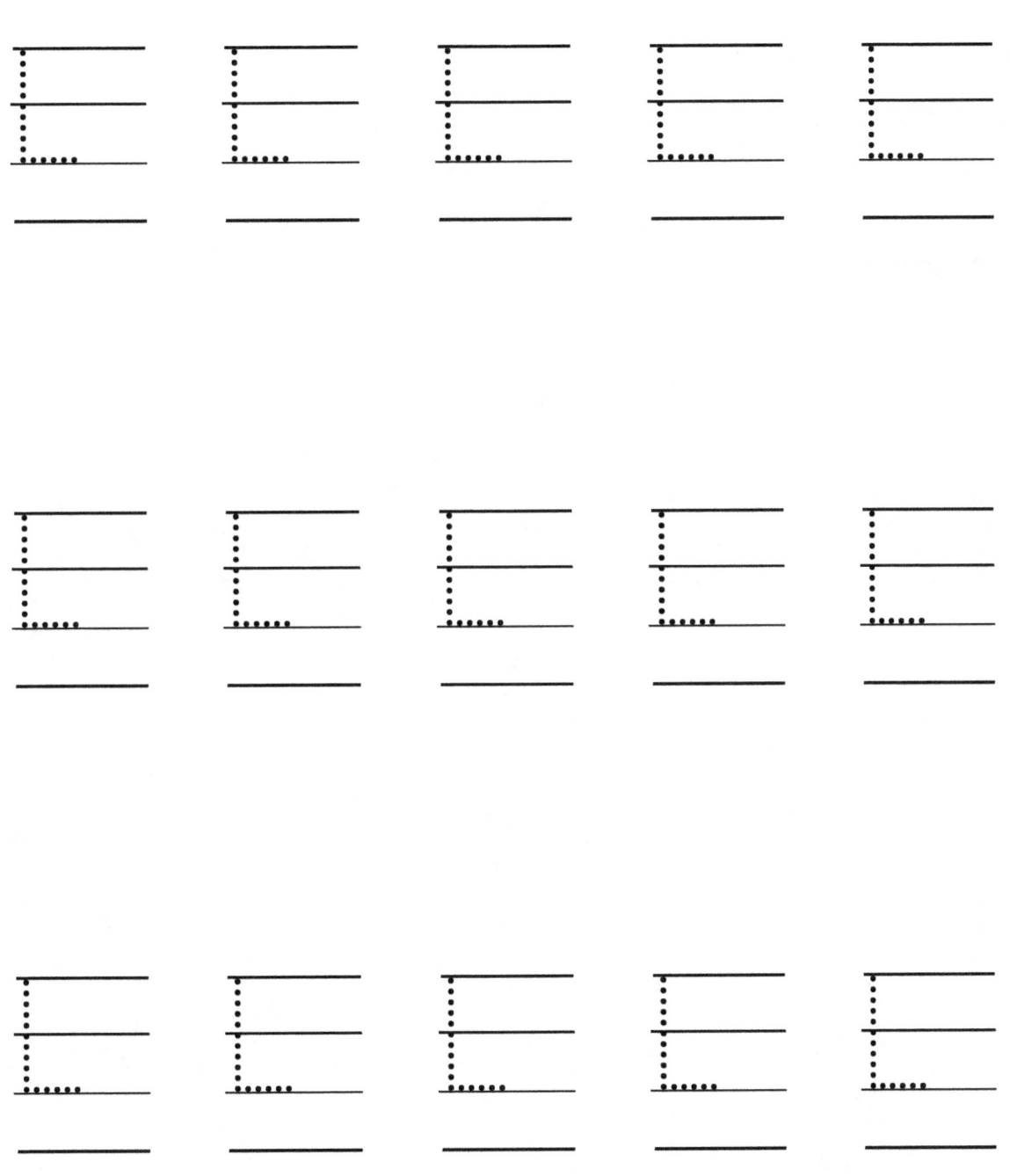

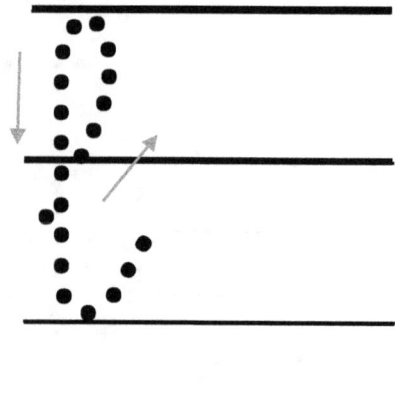

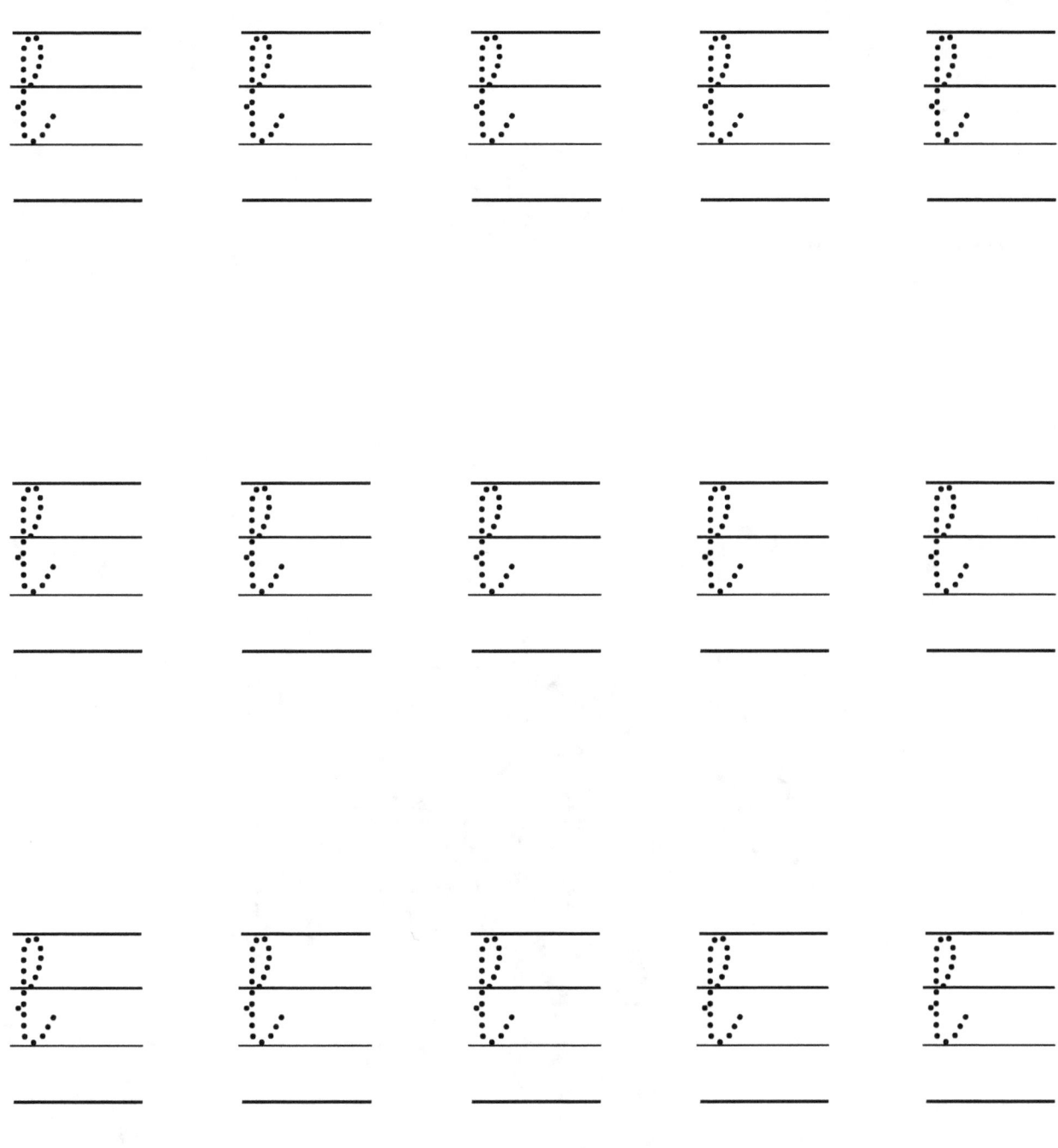

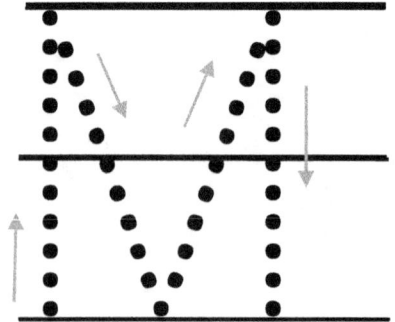

monkey

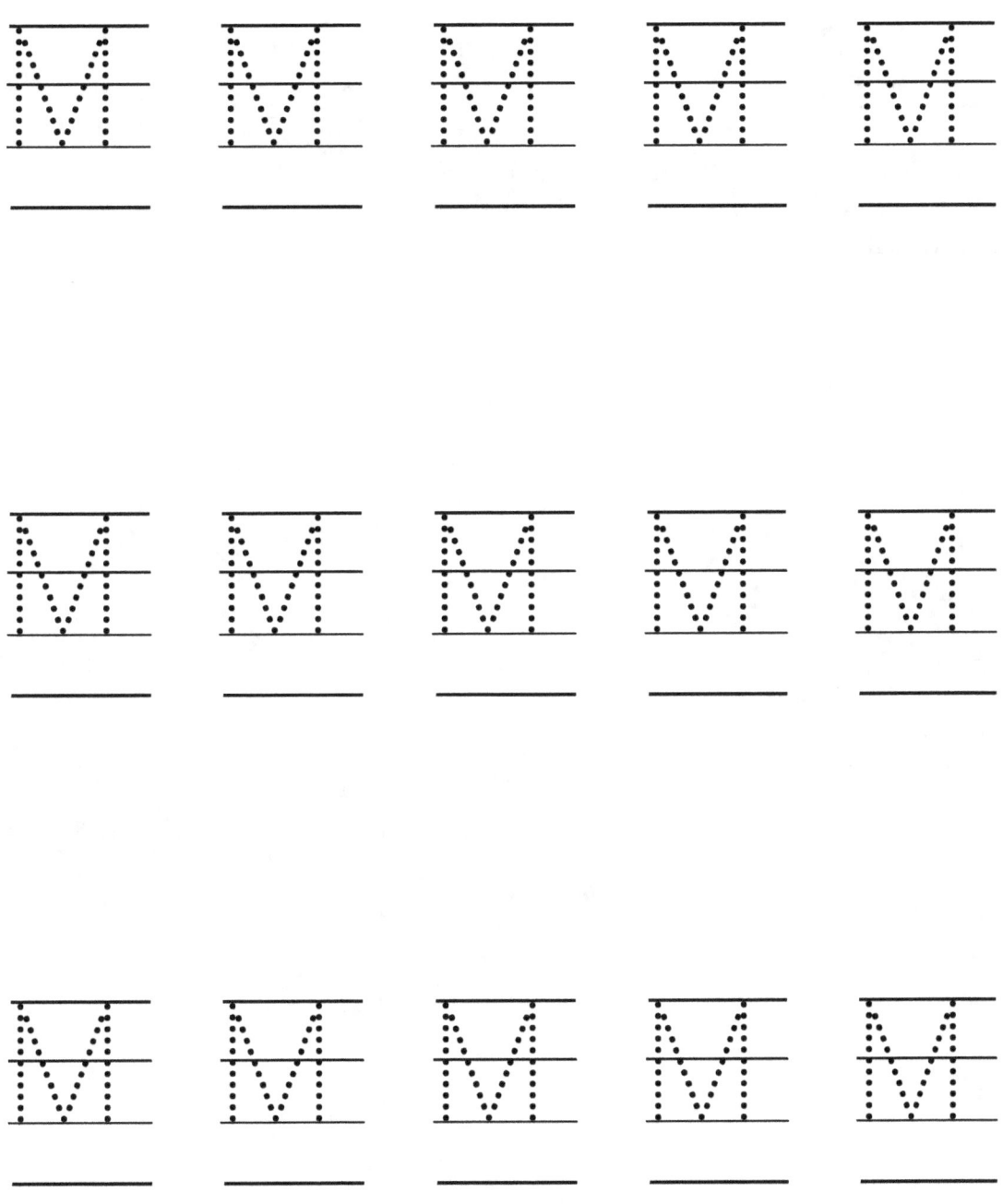

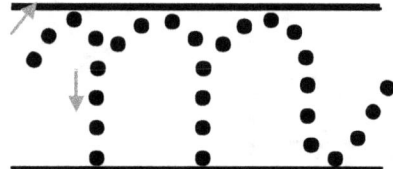

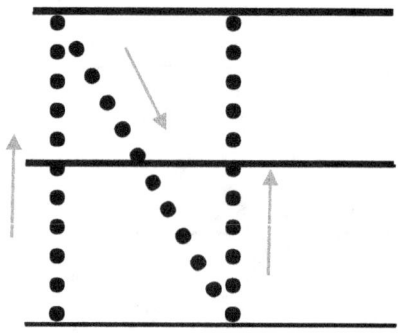

N nightingale

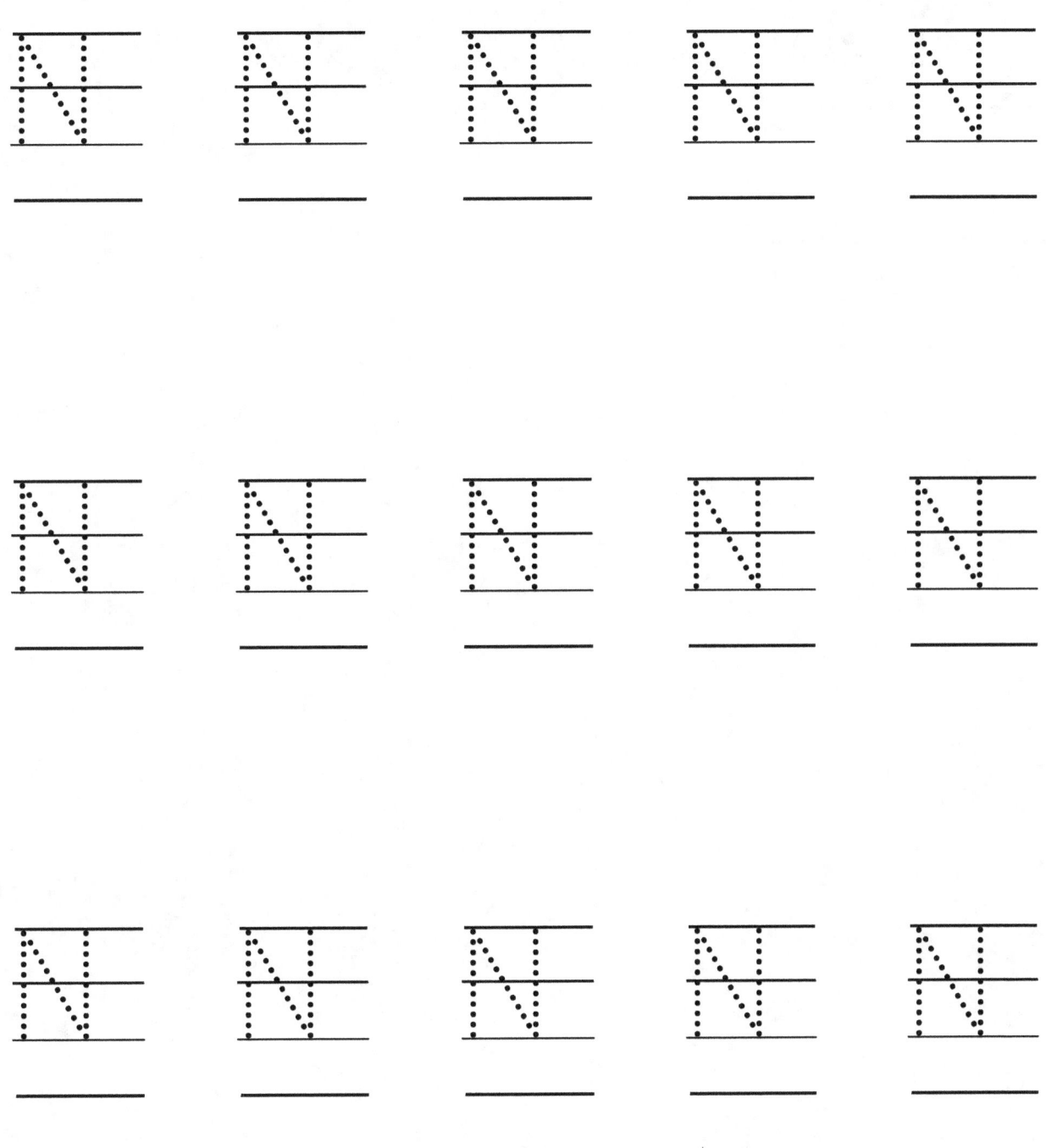

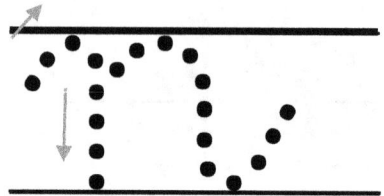

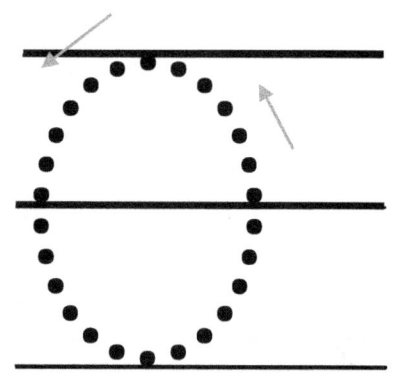

owl

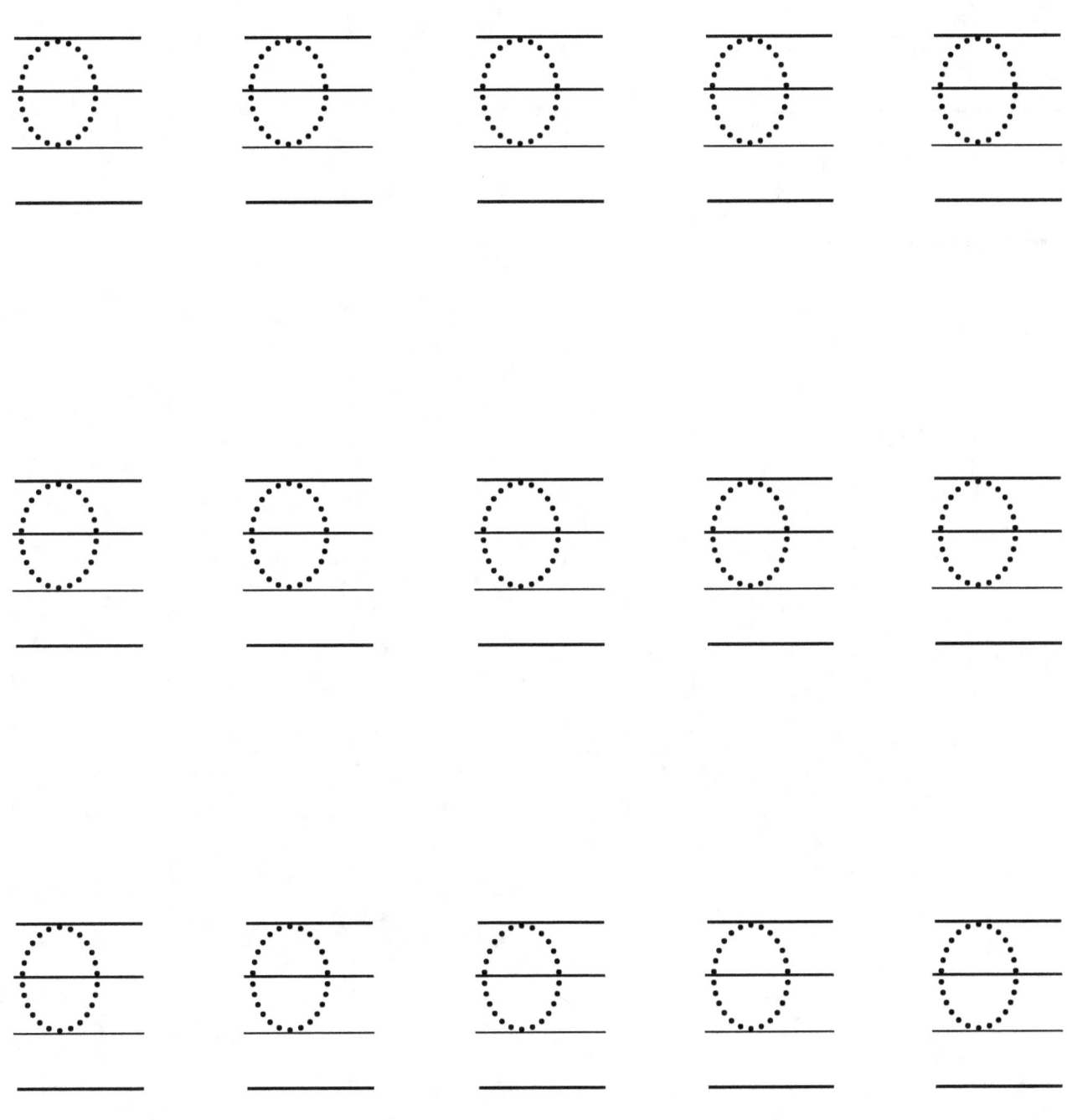

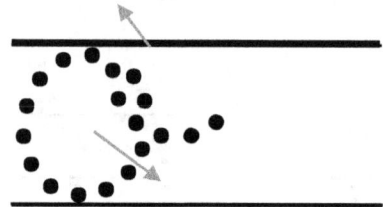

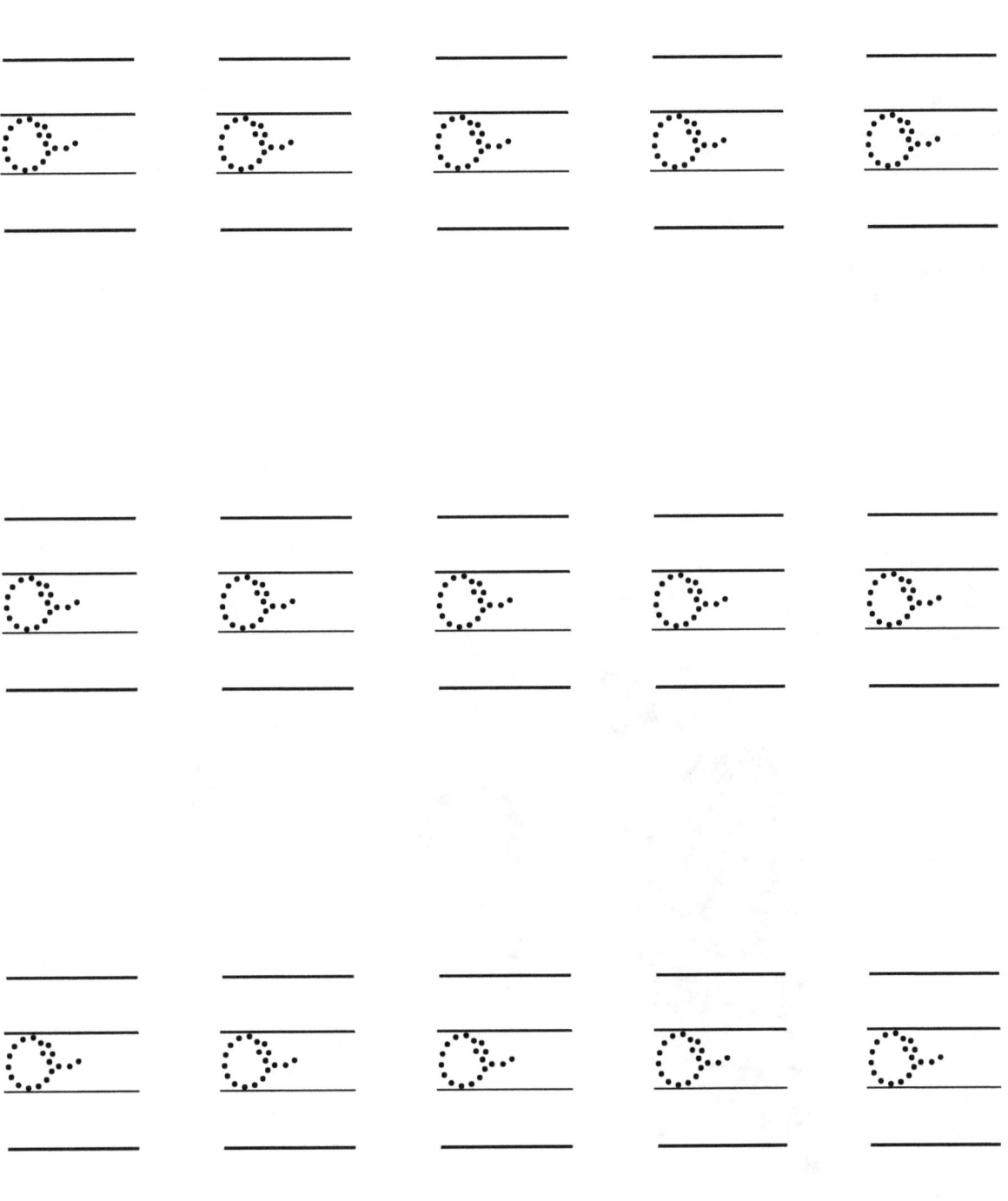

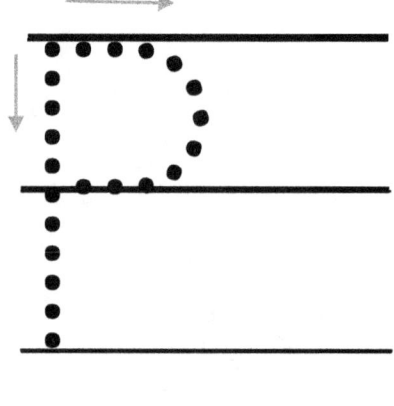

penguin

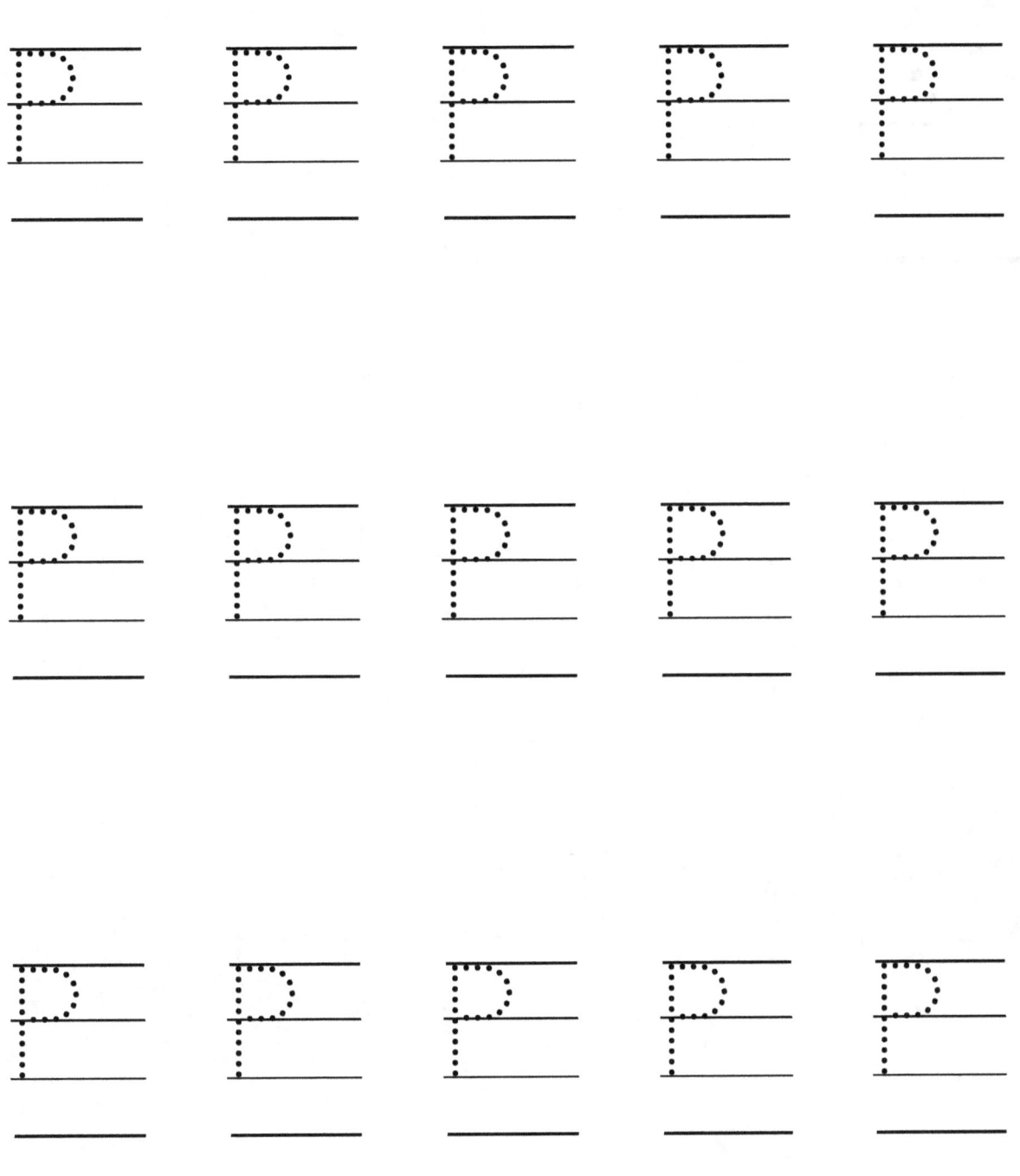

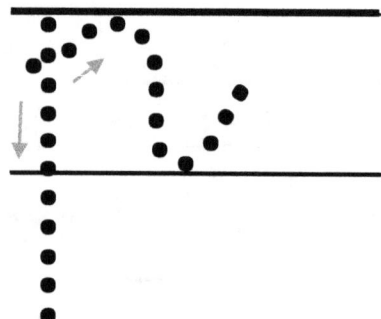

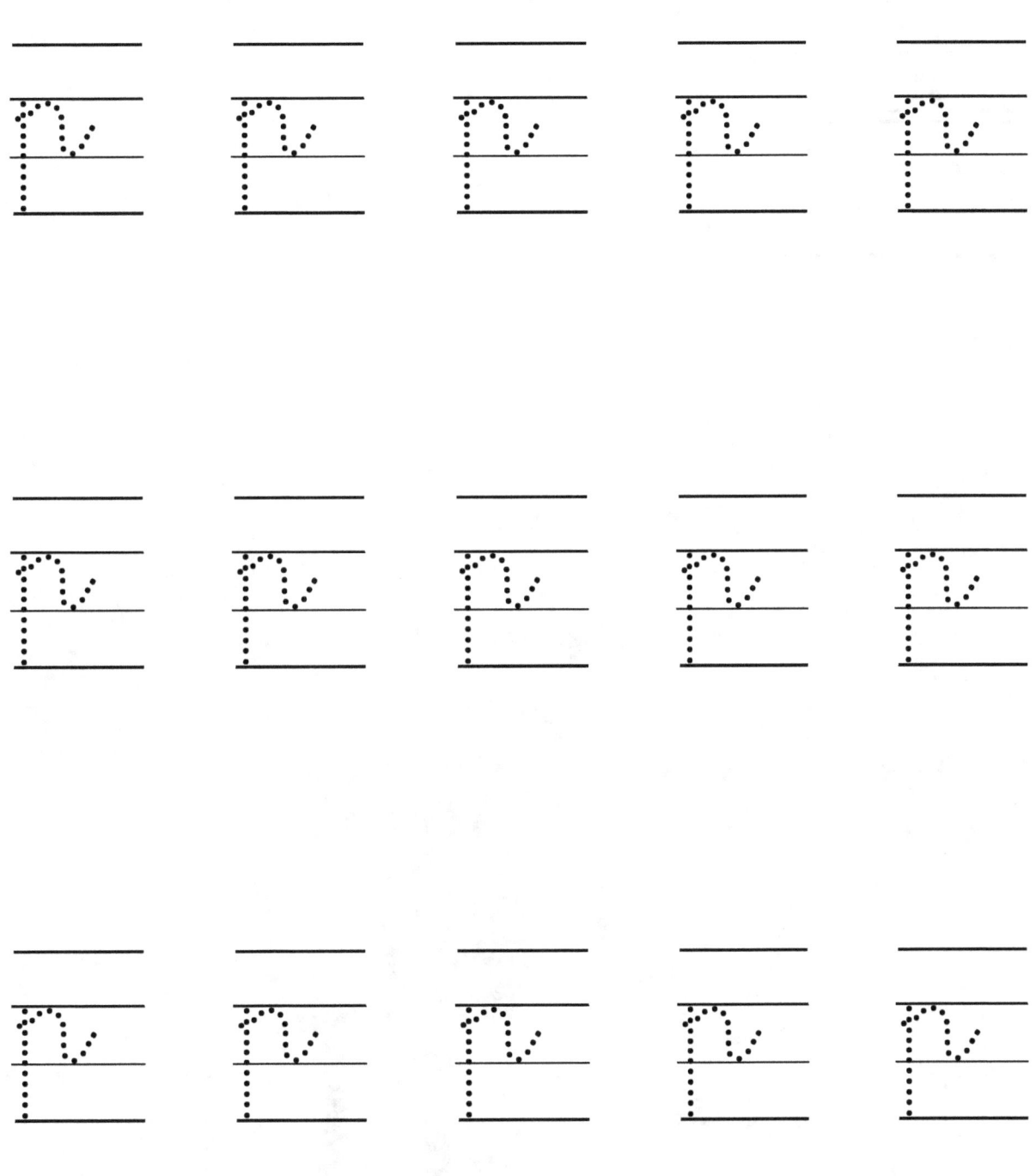

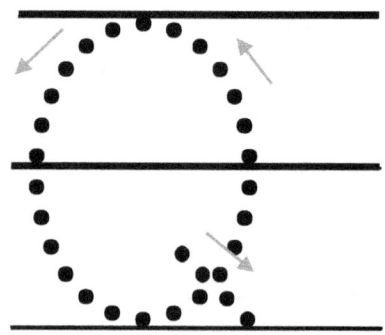

quail

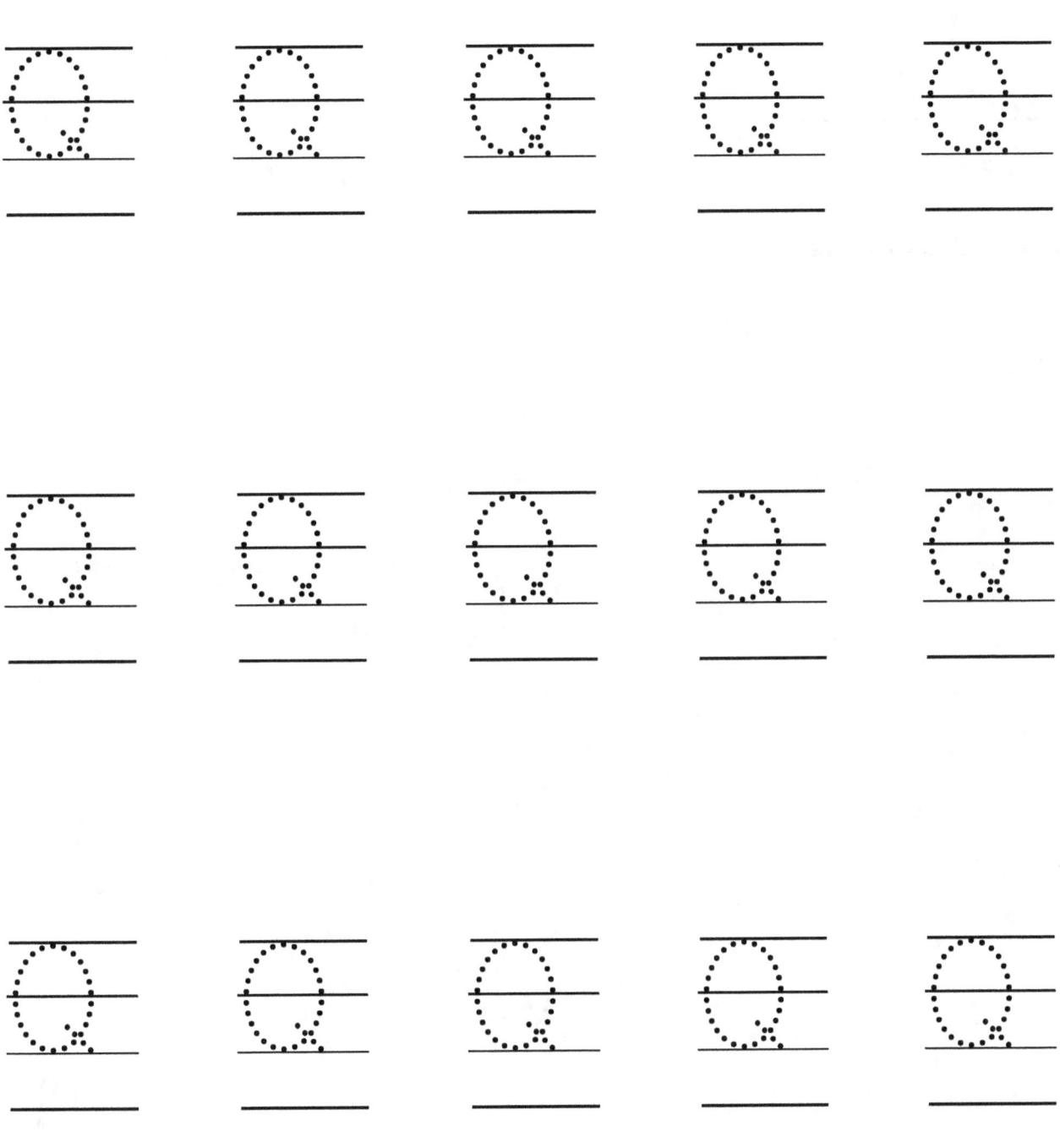

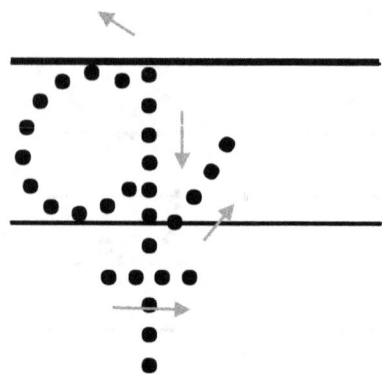

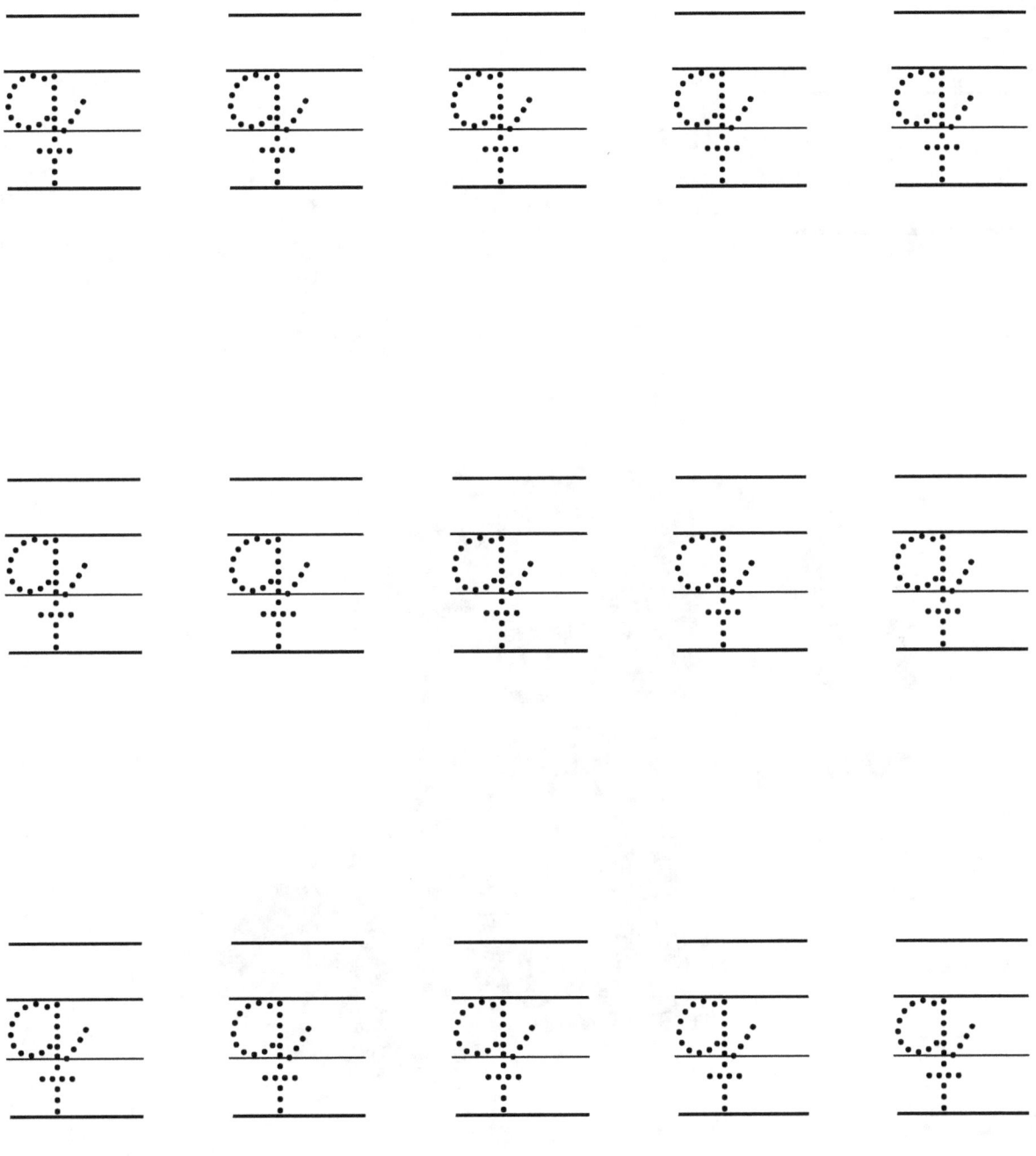

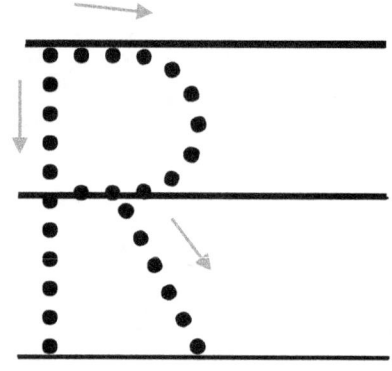

R
raccoon

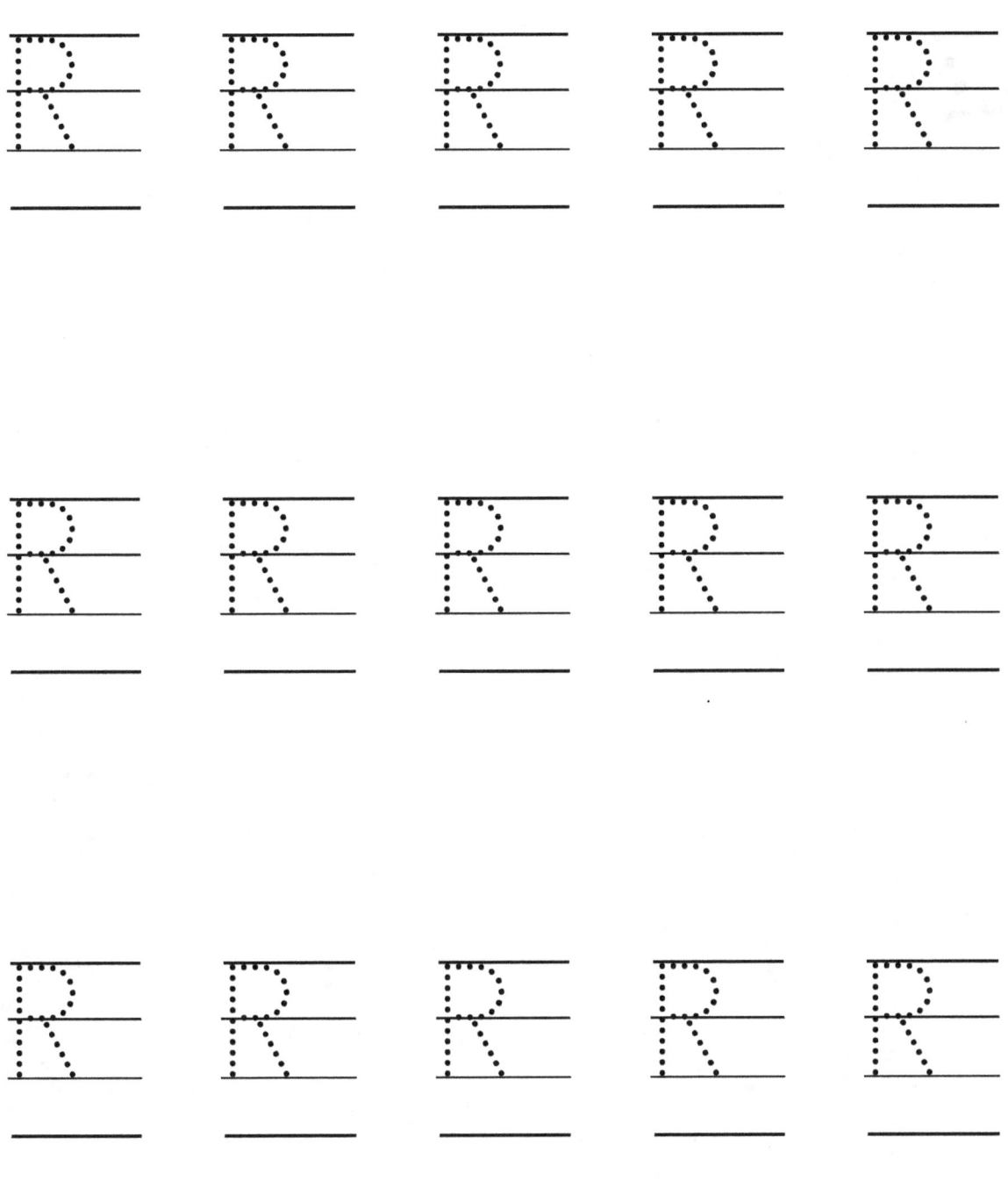

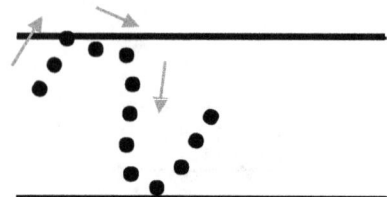

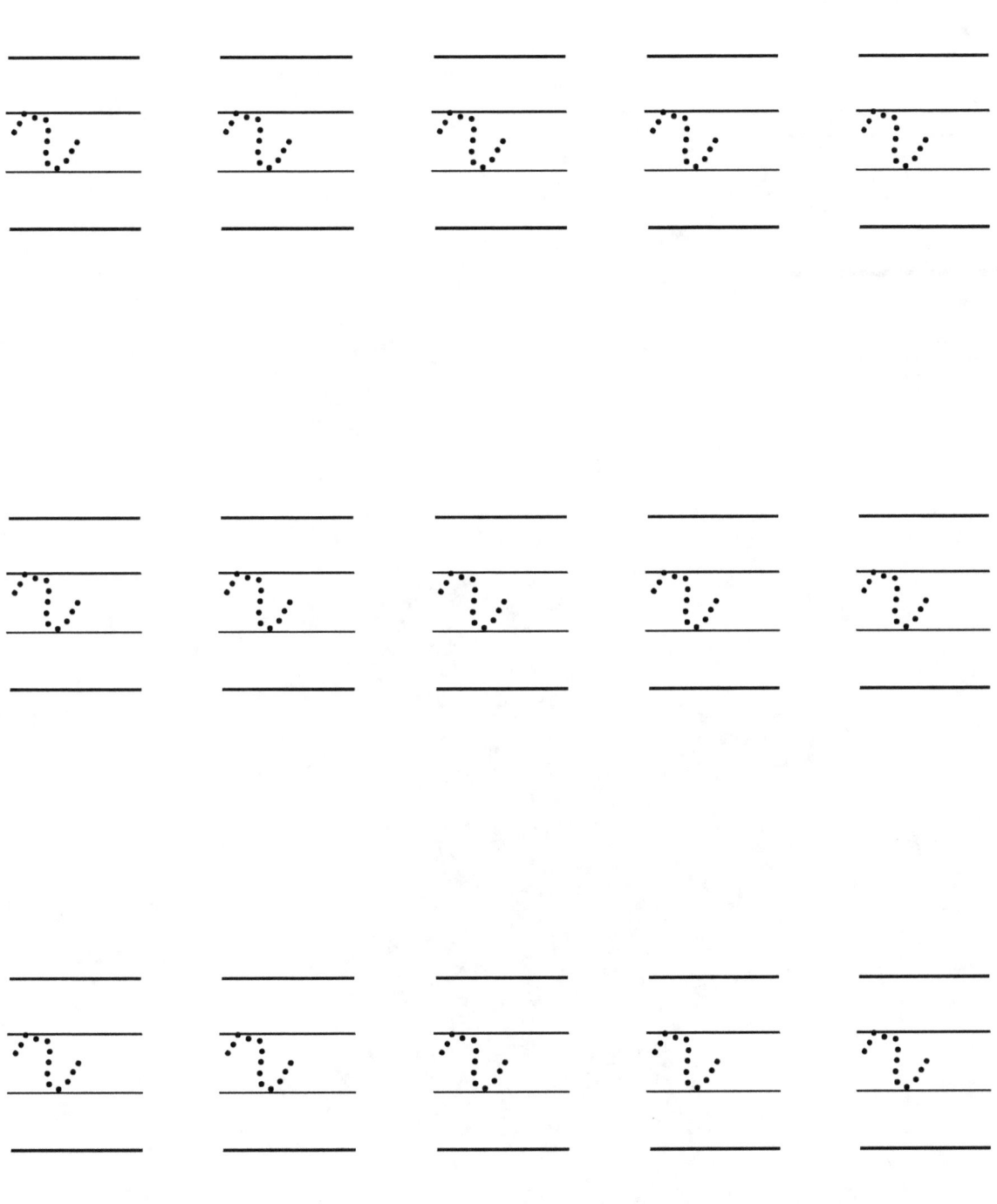

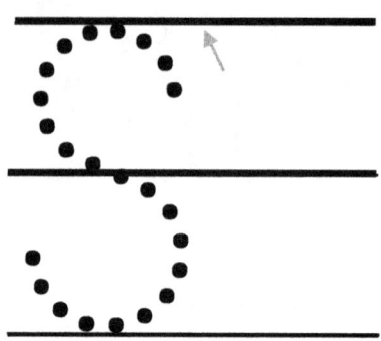

seal

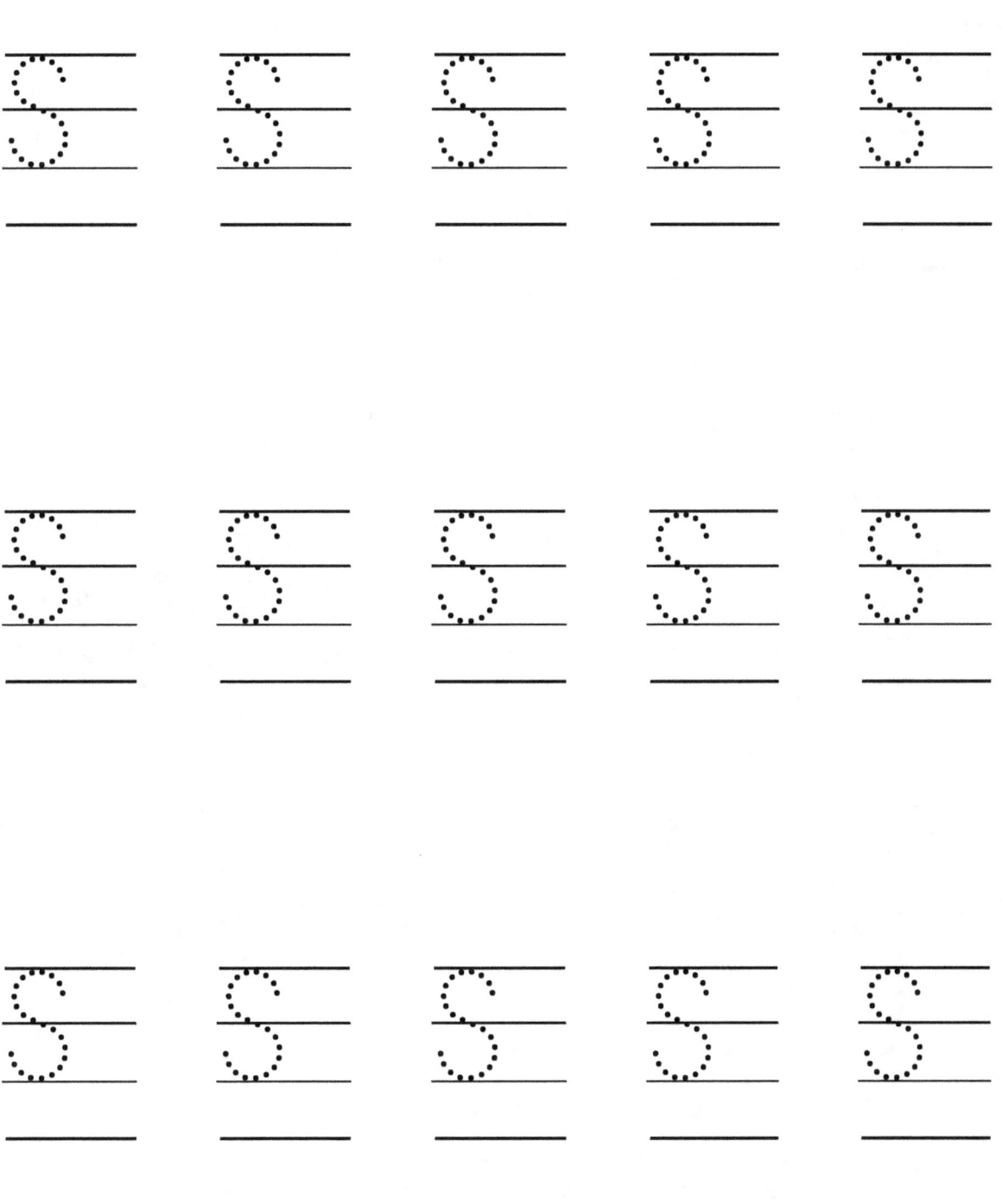

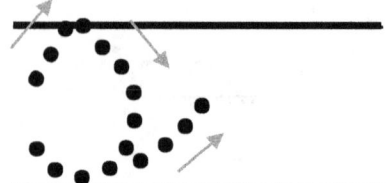

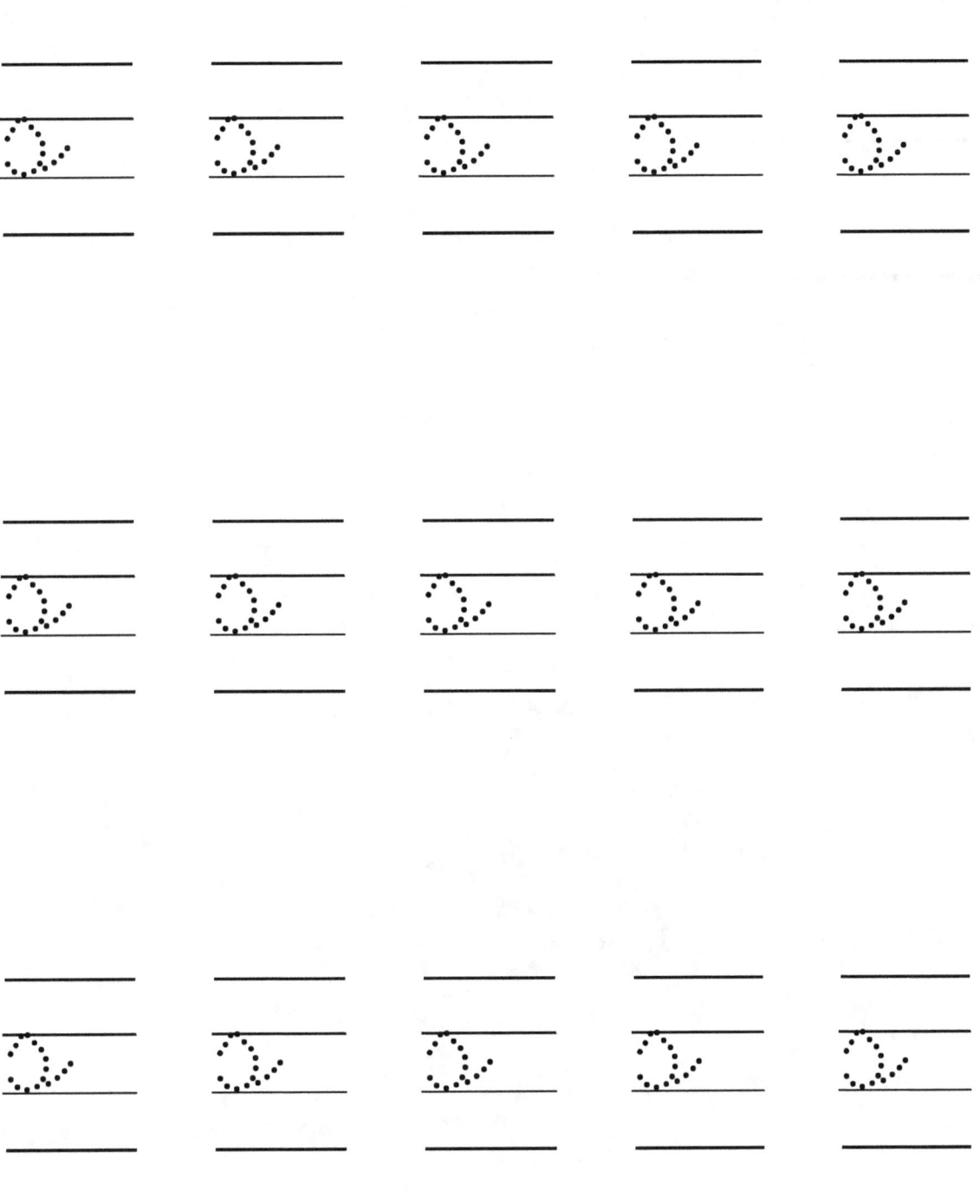

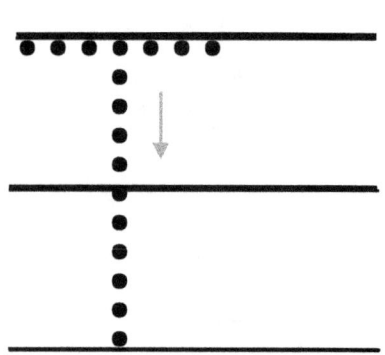

turtle

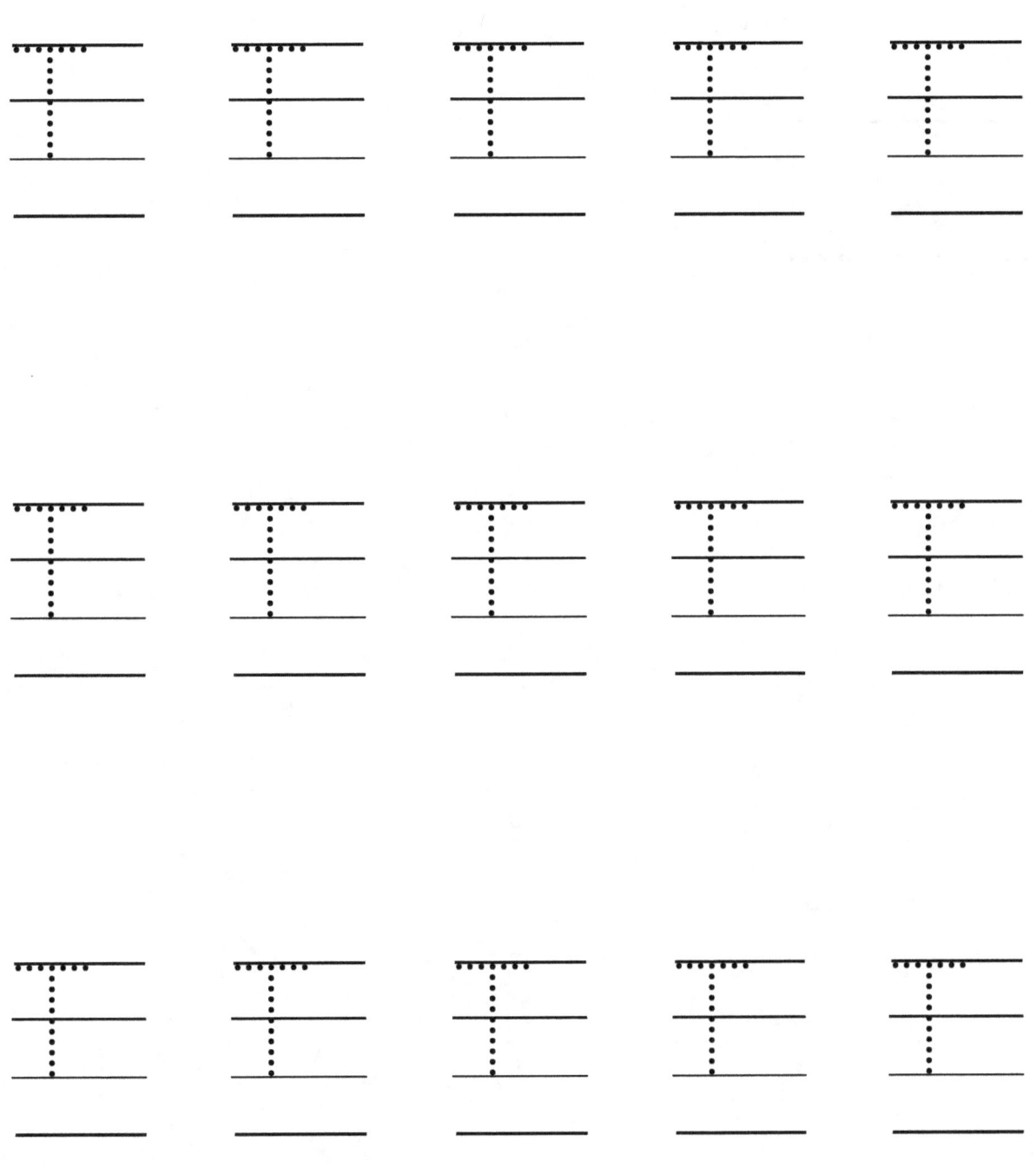

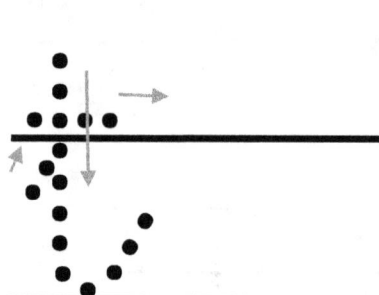

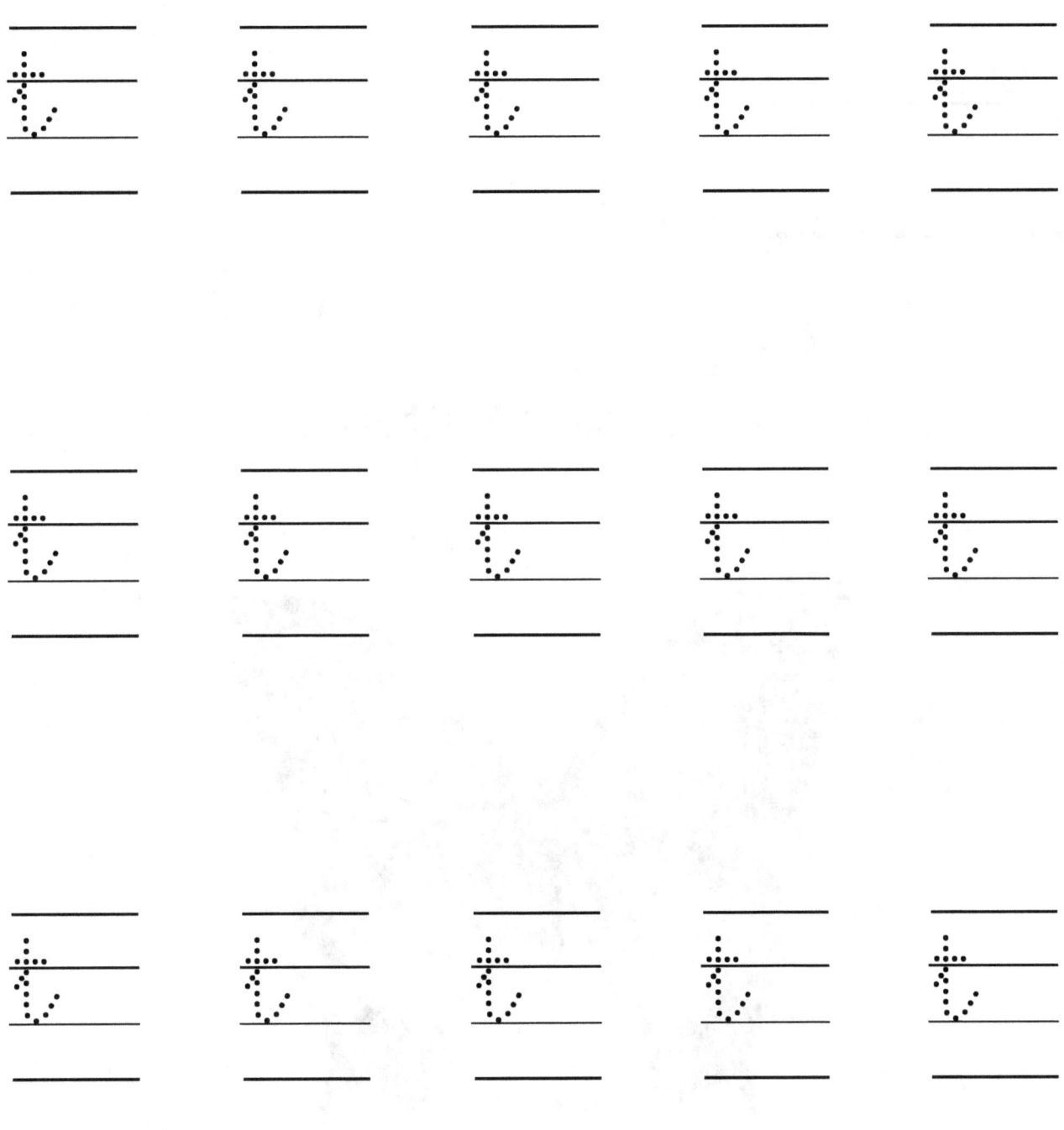

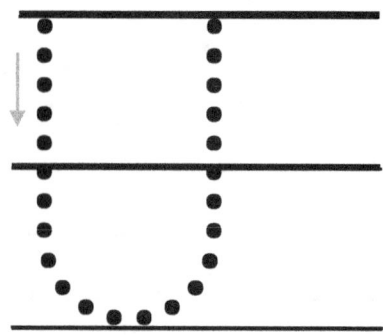

U Ulysses Butterfly

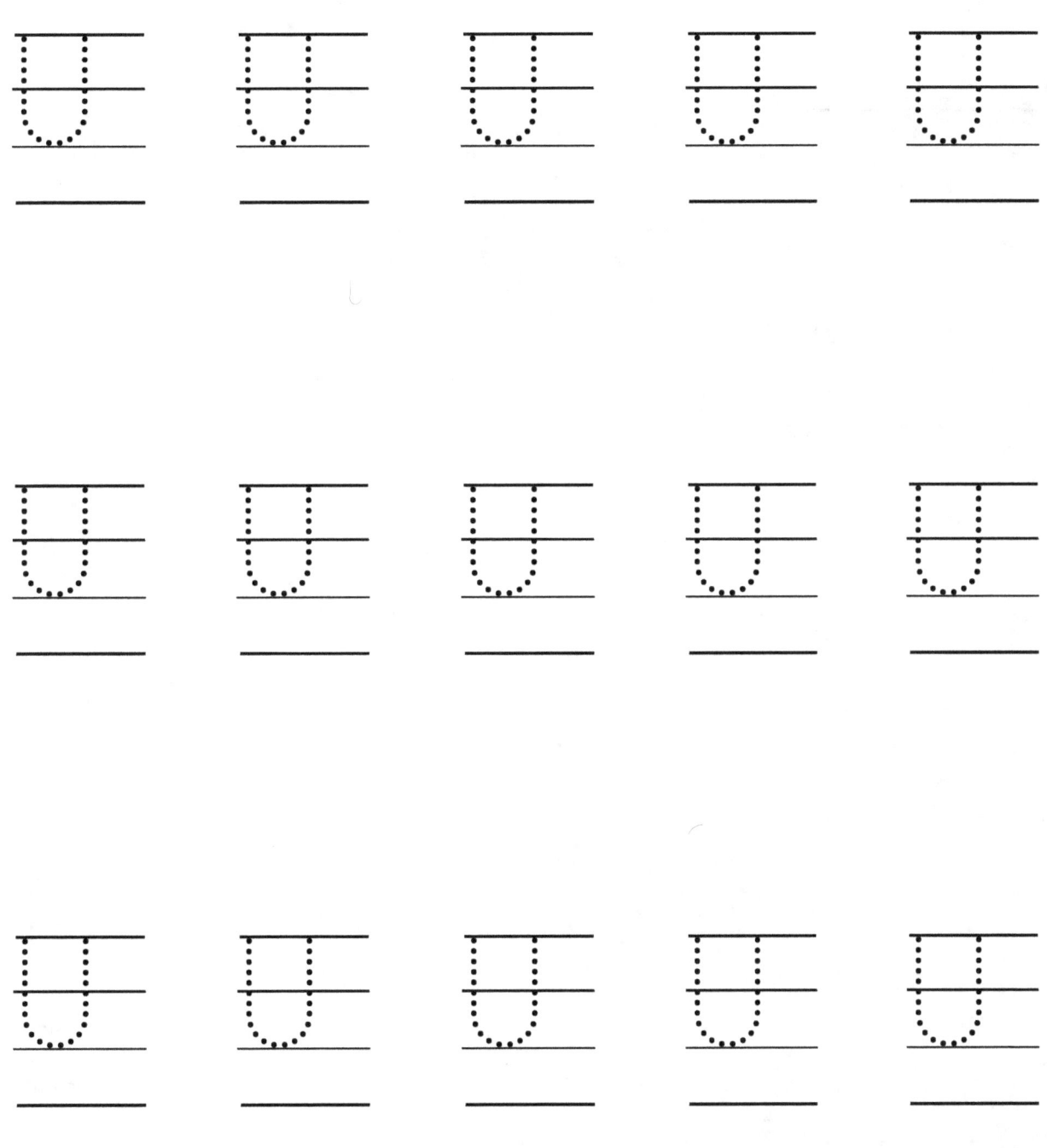

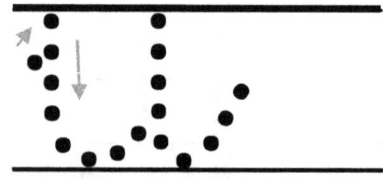

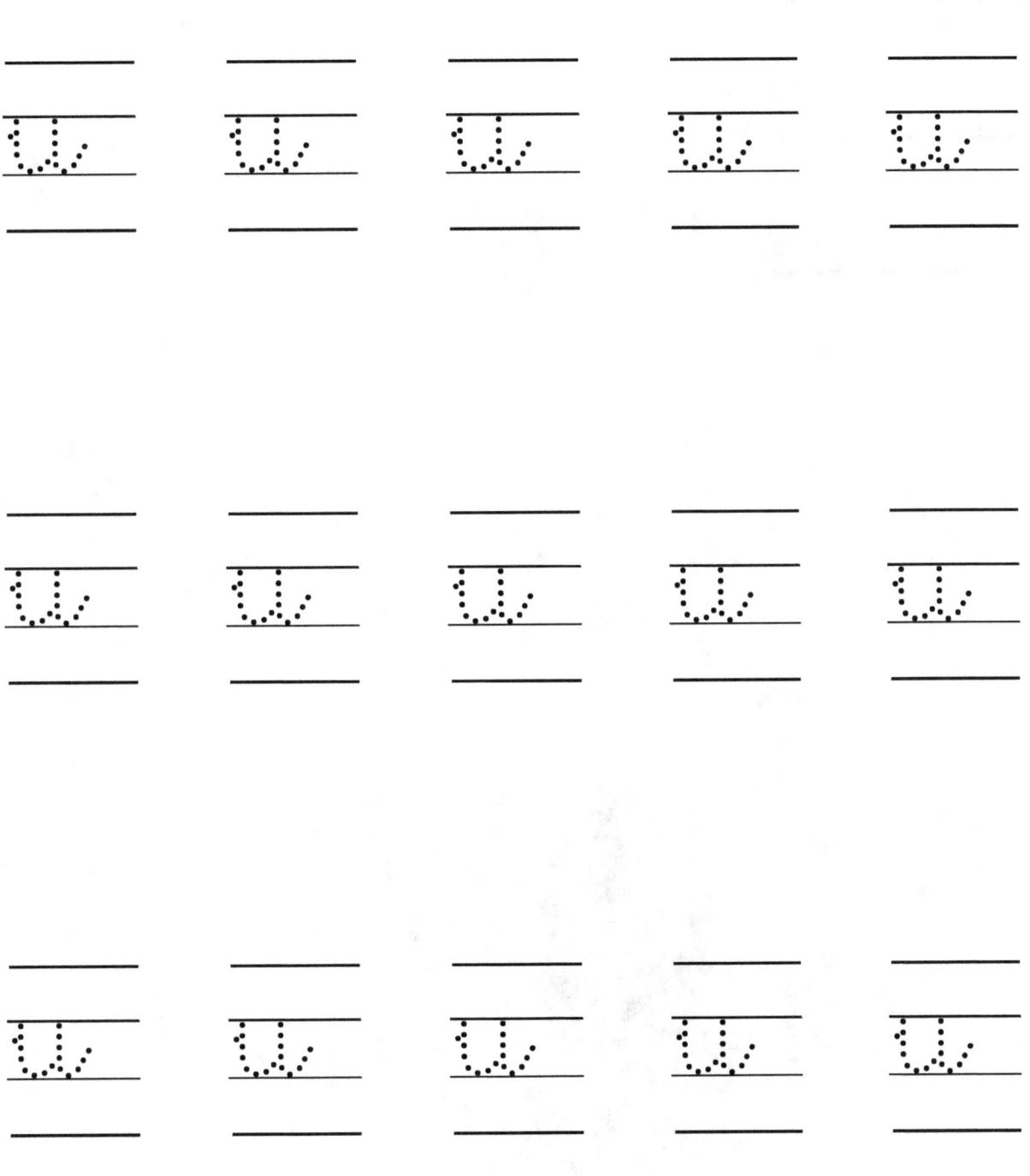

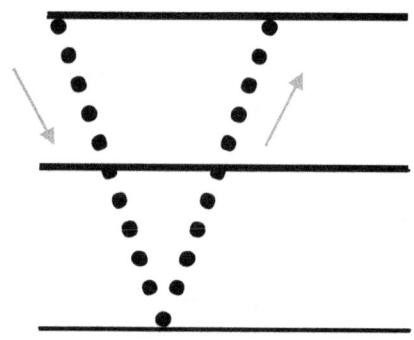

vulture

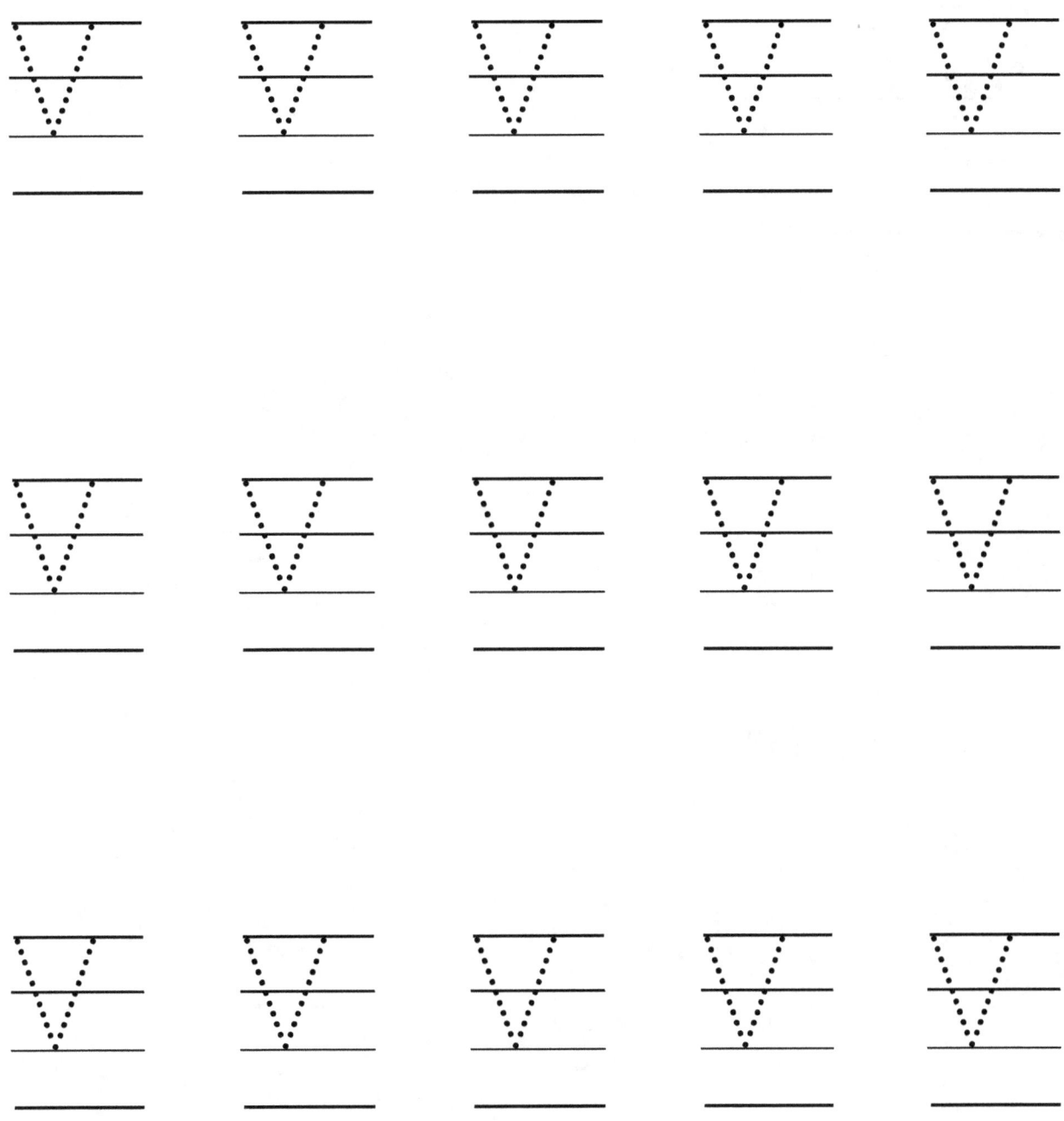

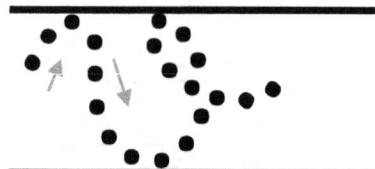

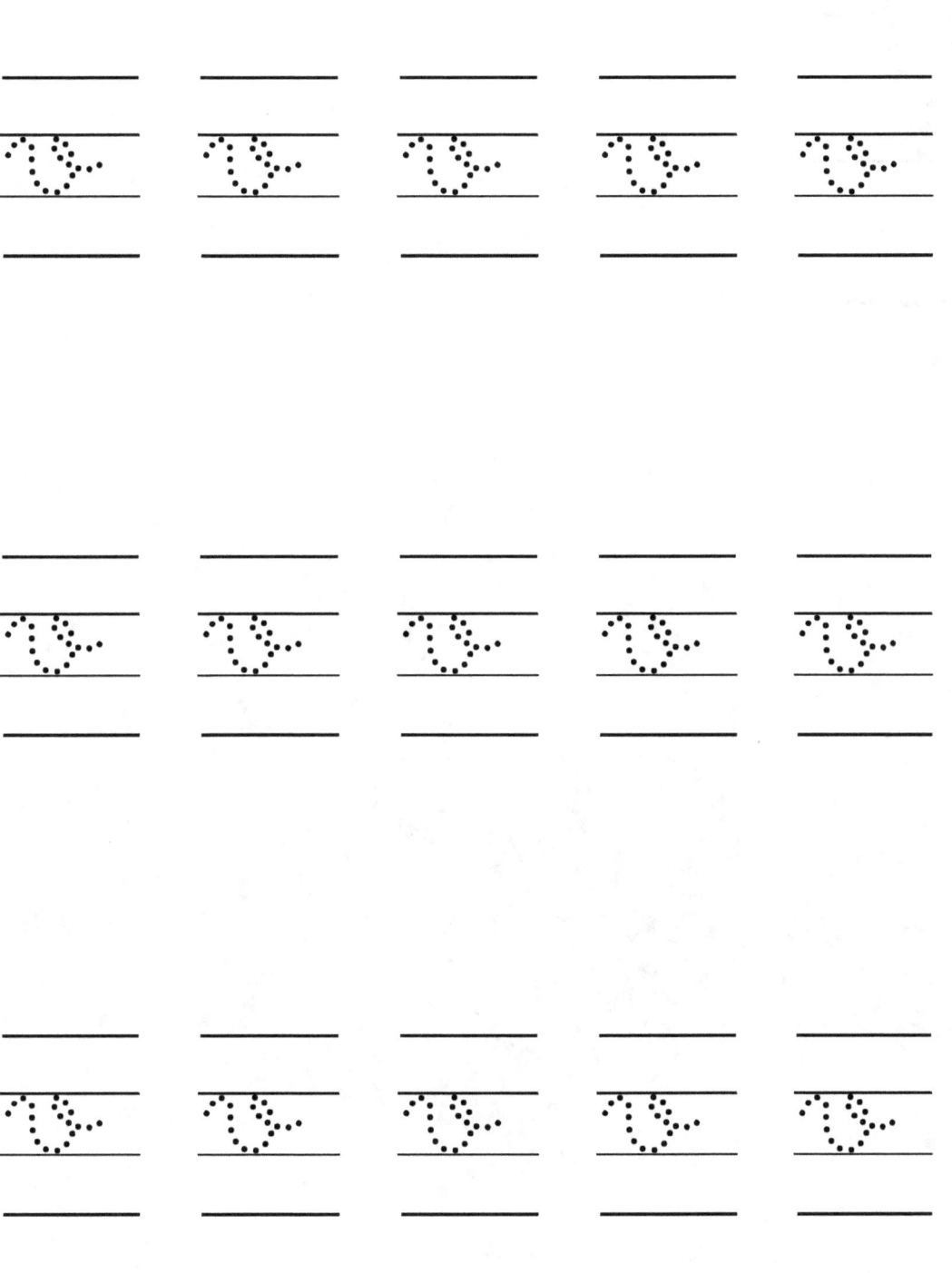

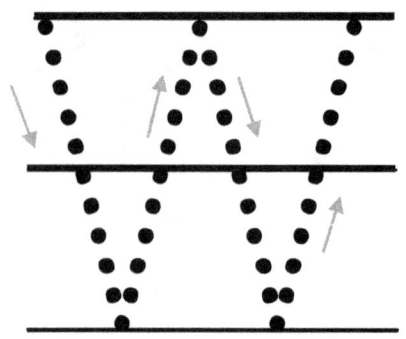

whale

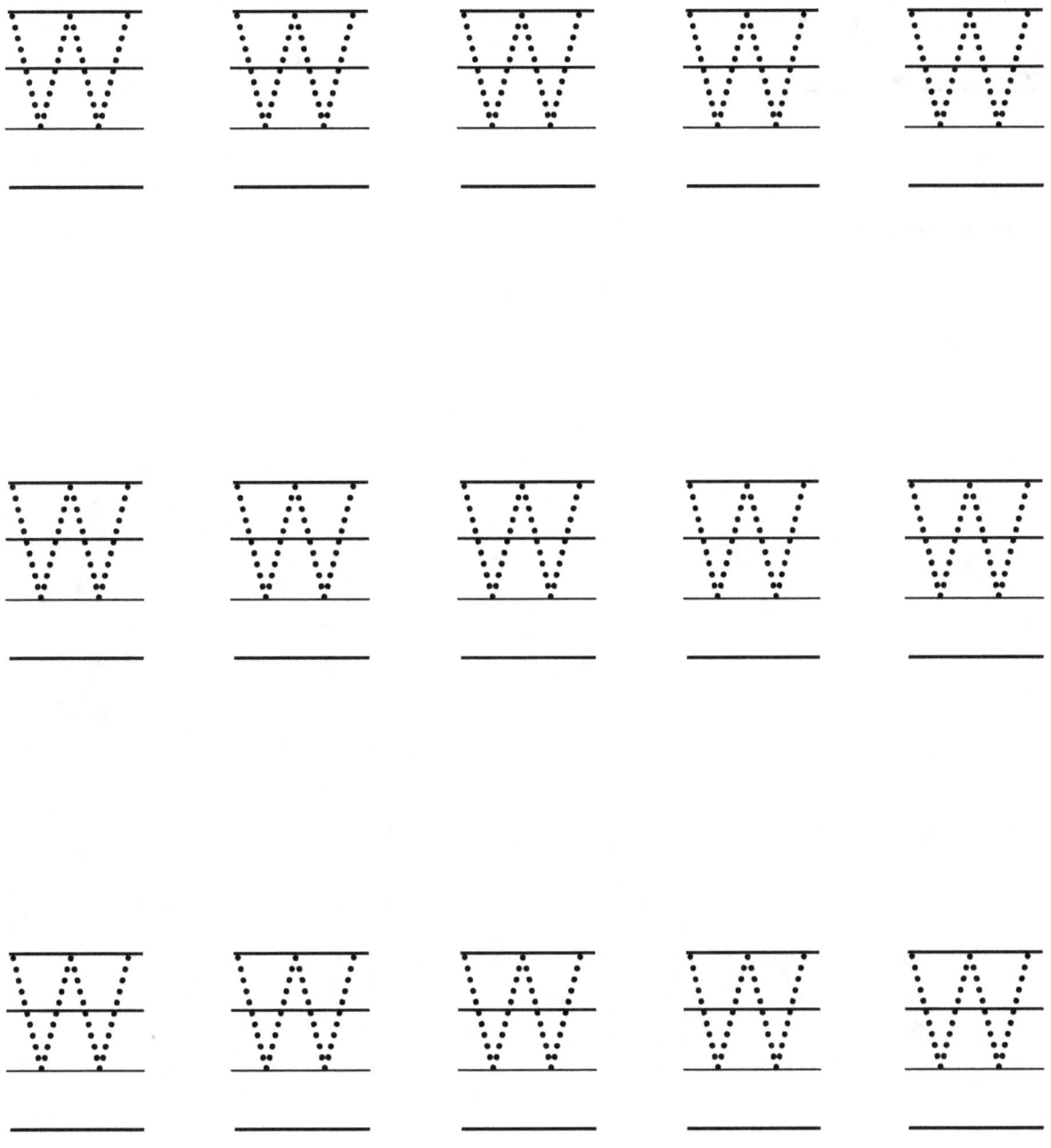

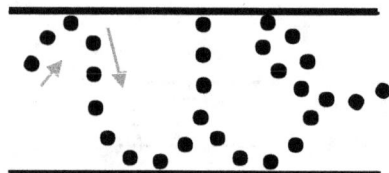

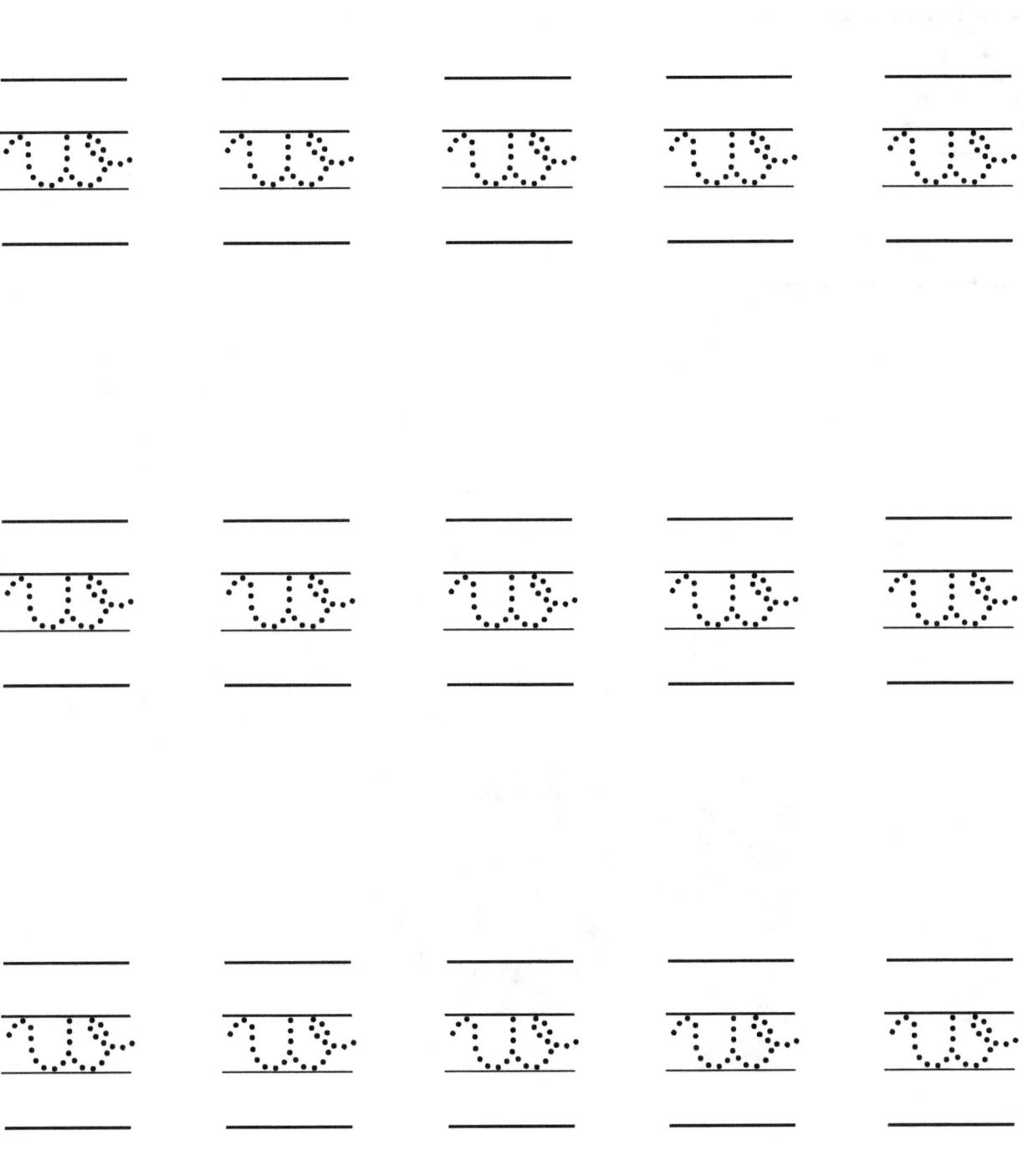

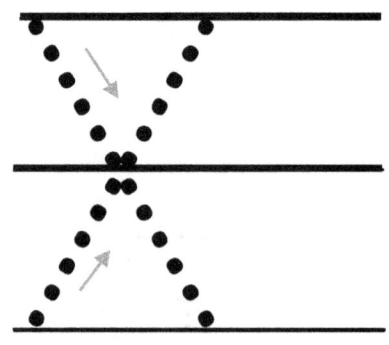

x-ray fish

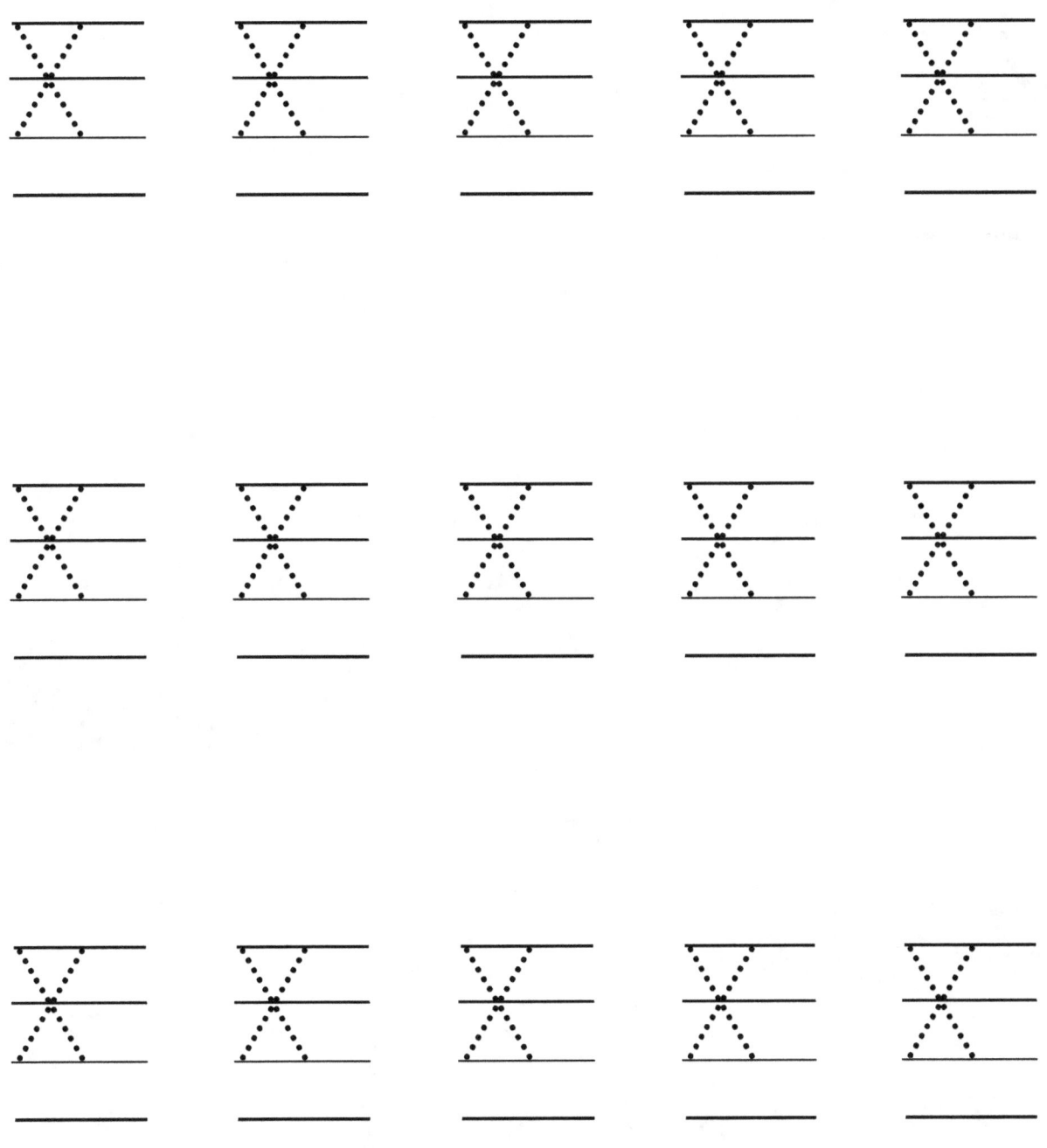

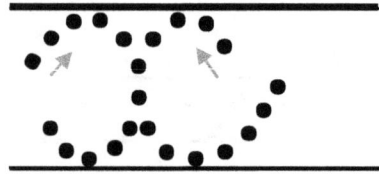

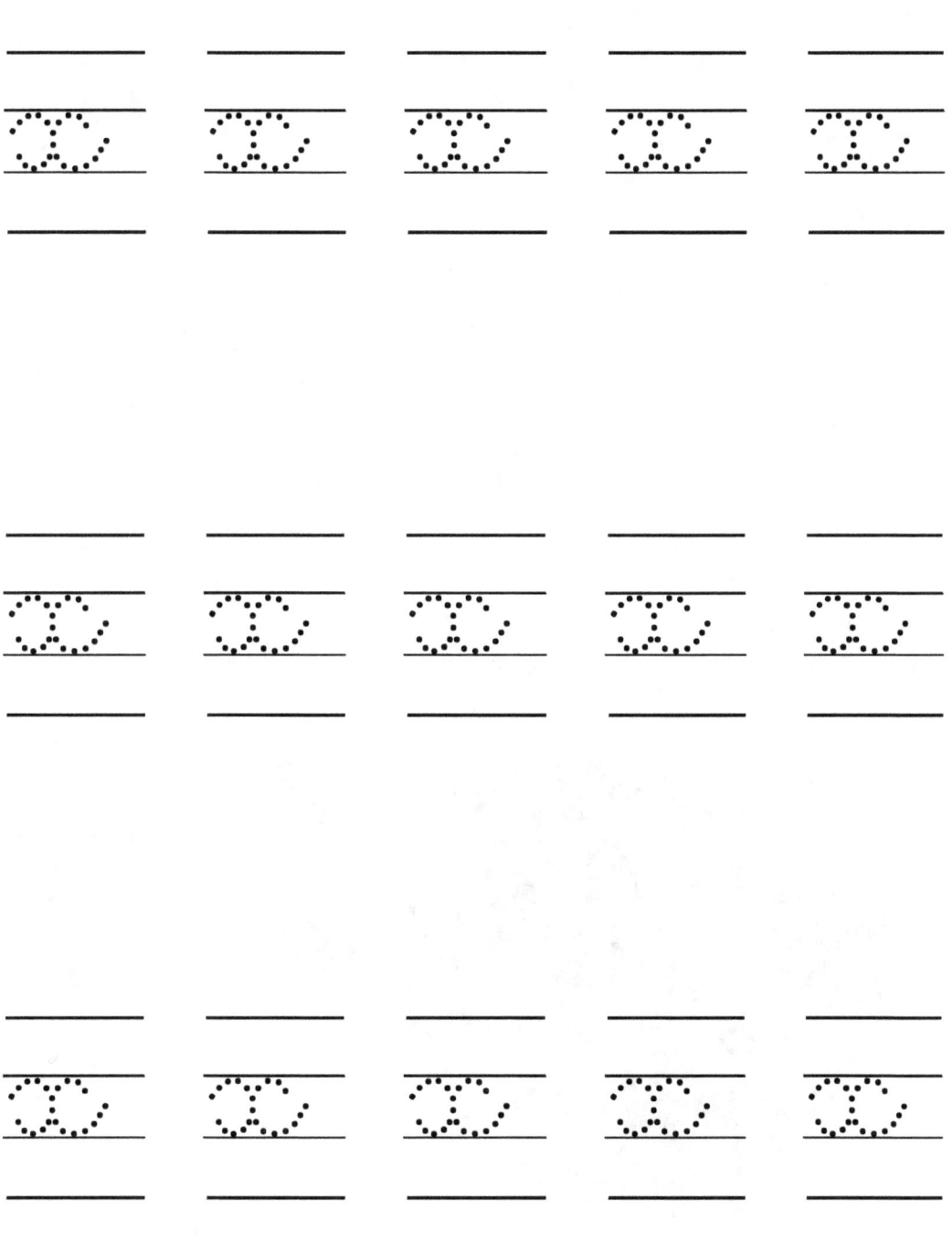

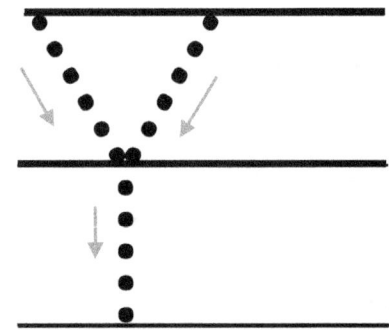

yak

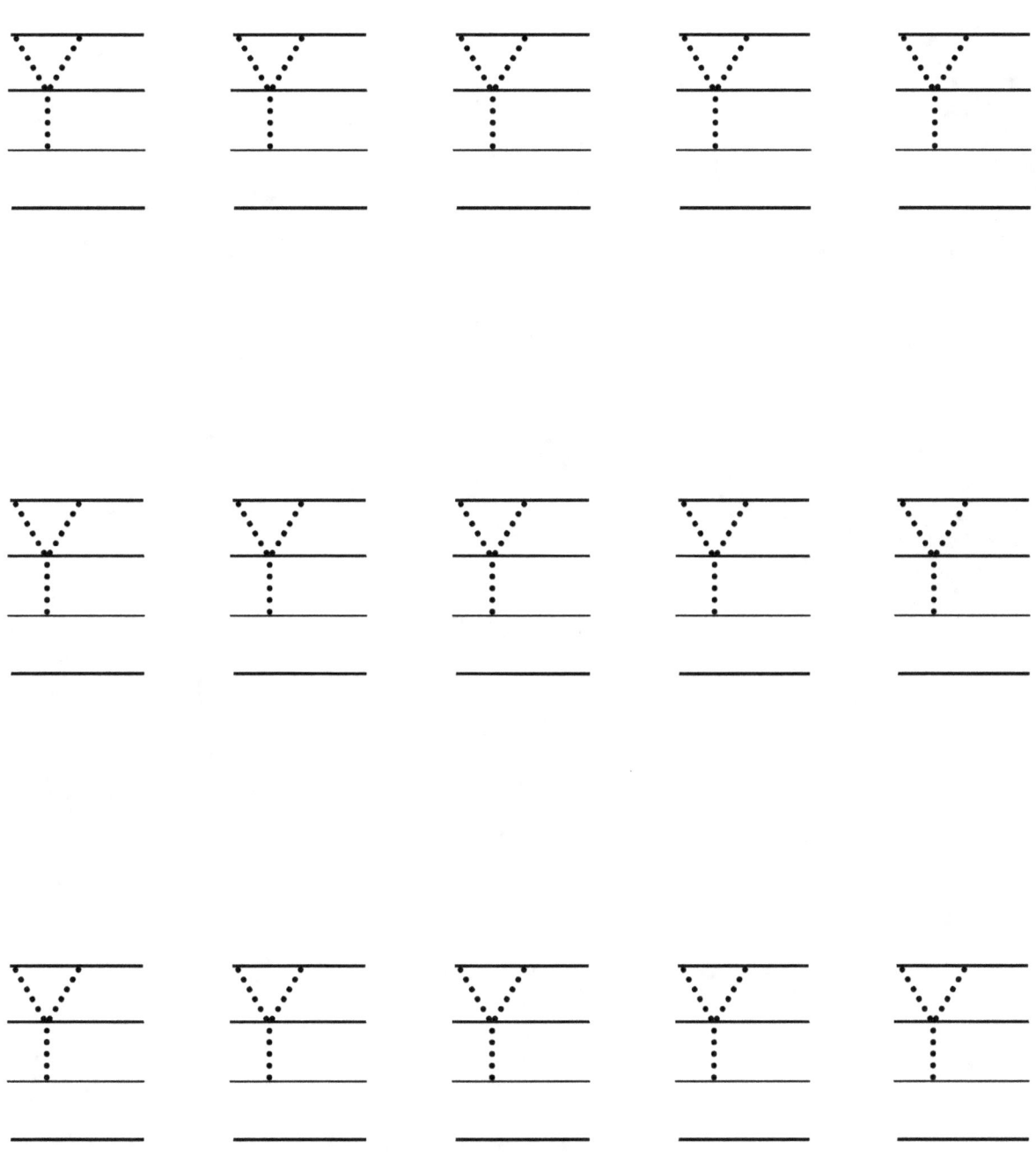

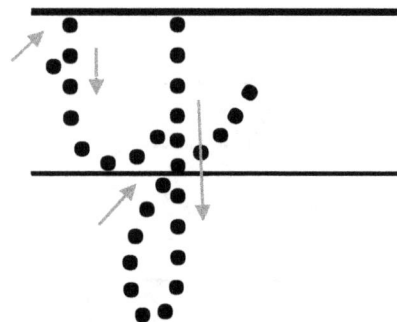

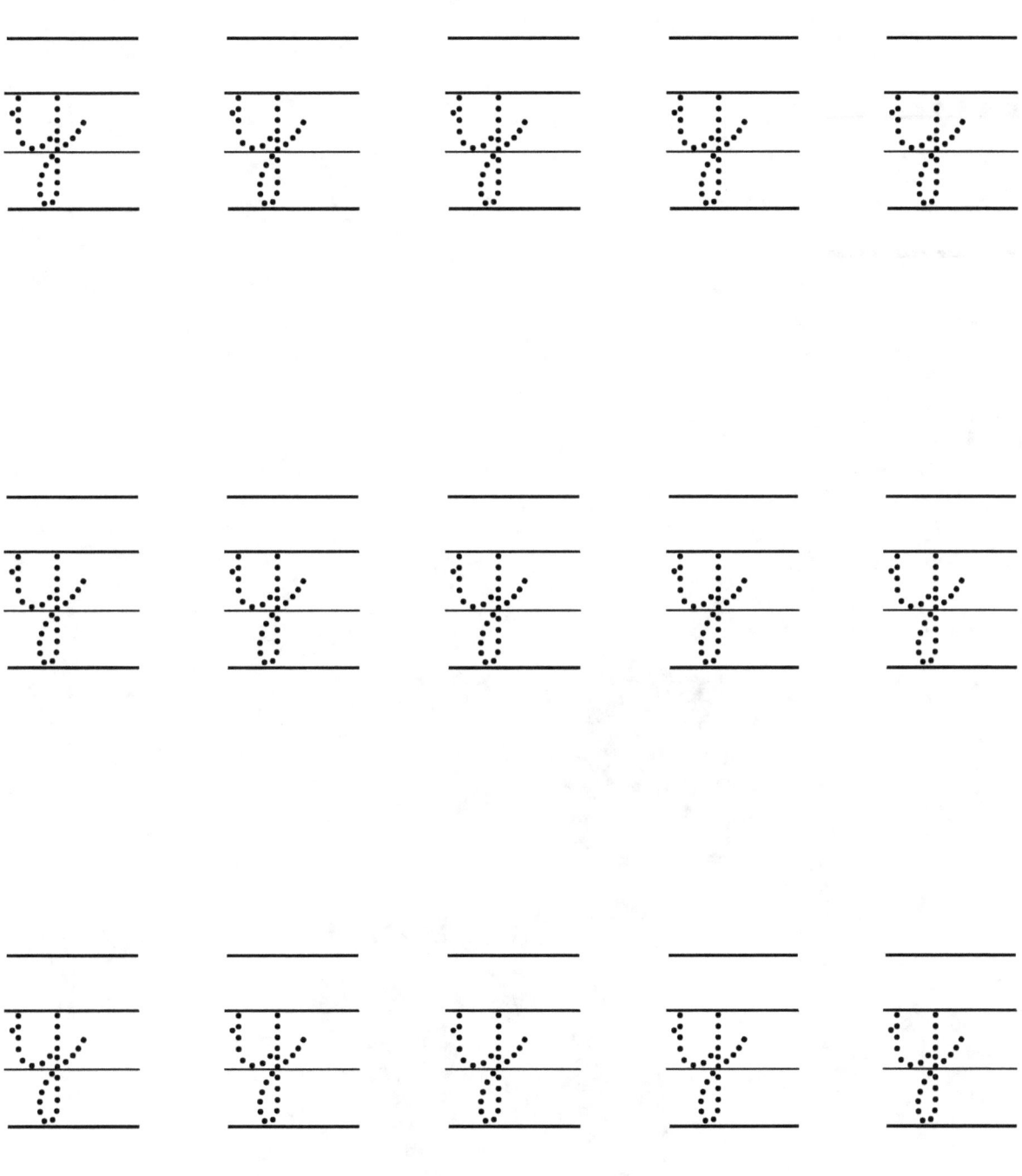

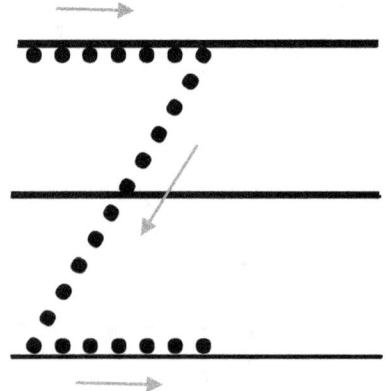

zebra

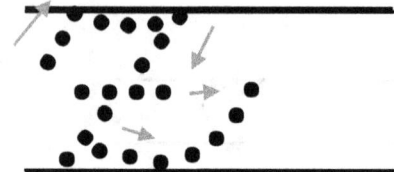

www.ingramcontent.com/pod-product-compliance
Lightning Source LLC
Chambersburg PA
CBHW081007170526
45158CB00010B/2949